COMPUTER AIDED
DESIGN AND MANUFACTURE

ELLIS HORWOOD SERIES IN ENGINEERING SCIENCE
Editors:
Prof. John M. Alexander, *Head of Dept. of Mechanical Engineering*
University College, Swansea
Dr. John Munro, *Reader in Civil Engineering*
Imperial College of Science and Technology, University of London
Prof. William Johnson, *Professor of Mechanical Engineering, Cambridge*
and Prof. S. A. Tobias, *Chance Professor of Mechanical Engineering*
University of Birmingham

The Ellis Horwood Engineering Science Series has two objectives; of satisfying the requirements of post-graduate and mid-career education and of providing clear and modern texts for more basic undergraduate topics in the fields of civil and mechanical engineering. It is furthermore the editors' intention to include English translations of outstanding texts originally written in other languages, thereby introducing works of international merit to English language audiences.

STRENGTH OF MATERIALS
J. M. ALEXANDER, University College of Swansea
TECHNOLOGY OF ENGINEERING MANUFACTURE
J. M. ALEXANDER, R. C. BREWER, Imperial College of Science and Technology, University of London, J. R. CROOKALL, Cranfield Institute of Technology.
VIBRATION ANALYSIS AND CONTROL SYSTEM DYNAMICS
CHRISTOPHER BEARDS, Imperial College of Science and Technology, University of London.
COMPUTER AIDED DESIGN AND MANUFACTURE
C. B. BESANT, Imperial College of Science and Technology, University of London.
STRUCTURAL DESIGN AND SAFETY
D. I. BLOCKLEY, University of Bristol.
BASIC LUBRICATION THEORY 2nd Edition
ALASTAIR CAMERON, Imperial College of Science and Technology, University of London.
ADVANCED MECHANICS OF MATERIALS 2nd Edition
Sir HUGH FORD, F.R.S. and J. M. ALEXANDER, Imperial College of Science and Technology, University of London.
RIGID BODY DYNAMICS IN ENGINEERING
R. GOHAR, Imperial College of Science and Technology, University of London.
TECHNIQUES OF FINITE ELEMENTS
BRUCE M. IRONS, University of Calgary, and S. AHMAD, Bangladesh University of Engineering and Technology, Dacca.
STRUCTURAL DESIGN OF CABLE-SUSPENDED ROOFS
L. KOLLAR, City Planning Office, Budapest and K. SZABO, Budapest Technical University.
FLUID POWER SYSTEM DYNAMICS AND CONTROL
D. McCLOY, The Northern Ireland Polytechnic and H. R. MARTIN, University of Waterloo, Ontario, Canada.
STATISTICS AND PROBABILITY IN ENGINEERING DECISION MAKING
J. MUNRO and P. W. JOWITT, Imperial College of Science and Technology, University of London.
DYNAMICS OF MECHANICAL SYSTEMS 2nd Edition
J. M. PRENTIS, University of Cambridge.
ENERGY METHODS IN VIBRATION ANALYSIS
T. H. RICHARDS, University of Aston, Birmingham.
ENERGY METHODS IN STRESS ANALYSIS: With an Introduction to Finite Element Techniques
T. H. RICHARDS, University of Aston, Birmingham.
STRESS ANALYSIS OF POLYMERS 2nd Edition
J. G. WILLIAMS, Imperial College of Science and Technology, University of London.

COMPUTER AIDED DESIGN
AND MANUFACTURE

C. B. BESANT, B.Sc., Ph.D., D.I.C.
Department of Mechanical Engineering
Imperial College of Science and Technology
University of London

ELLIS HORWOOD LIMITED
Publishers   Chichester

Halsted Press: a division of
JOHN WILEY & SONS
New York - Chichester - Brisbane - Toronto

First published in 1980 by
ELLIS HORWOOD LIMITED
Market Cross House, Cooper Street, Chichester, West Sussex, PO19 1EB, England

The publisher's colophon is reproduced from James Gillison's drawing of the ancient Market Cross, Chichester.

Distributors:

Australia, New Zealand, South-east Asia:
Jacaranda-Wiley Ltd., Jacaranda Press,
JOHN WILEY & SONS INC.,
G.P.O. Box 859, Brisbane, Queensland 40001, Australia.

Canada:
JOHN WILEY & SONS CANADA LIMITED
22 Worcester Road, Rexdale, Ontario, Canada.

Europe, Africa:
JOHN WILEY & SONS LIMITED
Baffins Lane, Chichester, West Sussex, England.

North and South America and the rest of the world:
Halsted Press, a division of
JOHN WILEY & SONS
605 Third Avenue, New York, N.Y. 10016, U.S.A.

British Library Cataloguing in Publication Data
Besant, C B
    Computer aided design and manufacture. —
    (Ellis Horwood series in mechanical engineering).
    1. Engineering design — Data processing
    I. Title
    620'.0042'02854   TA174   79-40971
ISBN 0-85312-117-6 (Ellis Horwood Ltd., Publishers, Library Edition)
ISBN 0-470-26868-9 (Halsted Press, Library Edition)

Typeset in Press Roman by Ellis Horwood Ltd.
Printed in Great Britain by W. & J. Mackay Ltd., Chatham

Table of Contents

Author's Preface

Computers are now beginning to be used extensively in the engineering industry for both design (CAD) and manufacture (CAM). Much of the industrial design is performed within the drawing office by design and production draughtsman. Many organisations have introduced CAD techniques into the drawing office in the form of automatic draughting systems or computing graphics systems. Further more, CAD and CAM techniques are now being taught in a large number of universities as part of engineering or computer science courses. At present there are very few books on CAD which meet the requirements of designer draughtsmen or engineering undergraduates who have had little computing experience.

This book aims to introduce the subject of computing as an aid to design and manufacture, and to take the reader through from the basics of computers to their application in real engineering draughting design and manufacture. It provides a description of both the hardware and software of CAD systems, together with a practical discussion of their use in engineering draughting. Two final chapters show how computer graphics, as part of the draughting process, can be linked with engineering analysis techniques to provide a CAD system for stressing engineering designs and for manufacturing engineering components by using NC machines.

Particular emphasis is given to the reader who has very little computing knowledge; and for those who become sufficiently interested to write their own graphics software, one chapter has been devoted to the description of algorithms used in computer graphics.

The book is based on the work of the CAD Section at Imperial College, and I am indebted to all who have worked with me in this field over the past six years. In particular I would mention some of my past research students, namely: Drs R. E. Grindley, A. D. Hamlyn, P. H. Huckle, D. Thompson, F. Ghassemi, and D. P. Craig, who have contributed so much to the success of the CAD and CAM research. Thanks are also due to the many firms who sponsored the CAD research and whose staff contributed so many ideas.

CHAPTER 1

Introduction

1.1 HISTORICAL INTRODUCTION

In the nineteenth century, the industrial revolution considerably enhanced man's physical powers. In the present century a second industrial revolution is taking place, with computers offering an enhancement of man's mental capabilities. Everyone is familiar with the use of computers in business, for example in banks for accounting and in companies for preparing lists of salaries and wages: many people realise the powerful contribution that computers are making in the world of science for performing large numbers of calculations. It is therefore surprising that the use of computers in engineering, particularly in the field of design, has so far not been properly understood or exploited.

The application of computers to various tasks often depends on the method of communication. In most business or scientific applications a teletype, similar to a conventional typewriter, is suitable as an input and output device to a computer. However, in engineering, particularly in design, the teletype on its own is not adequate as a communication device. The reason is that engineers traditionally communicate information graphically in the form of drawings. It can therefore be seen that visual communication through the medium of graphical representation is an essential part of engineering. However, computers find pictures difficult to handle, because large amounts of data are needed to describe most pictures in a form that a computer can interpret. This presents difficulties to engineers because the input of graphical information into a computer via a teletype can be time-consuming and tedious, and moreover is prone to errors.

The first important step forward in computer graphics was at Massachusetts Institute of Technology in 1963 when a system called SKETCHPAD [1] was demonstrated. This system consisted of a cathode-ray oscilloscope driven by a Lincoln TX2 computer whereby graphical information was displayed on the screen. Pictures could be drawn on the screen and then manipulated by using a device called a light pen. The use of systems based on the SKETCHPAD has become known as INTERACTIVE GRAPHICS. These systems are expensive because they are sophisticated and make demands on the power and resources of the computer. Therefore they were adopted only in such major industries as the aircraft industry where their use in design justifies the high capital costs.

There has been great progress in the development of cathode-ray oscilloscopes for the VISUAL DISPLAY UNIT (VDU) for use in computer graphics. The cost of such displays has become sufficiently low that computer graphics can now bring economic benefits to design in a wide range of industries.

A second area where important development studies have been made is in the computer itself. The minicomputer brought about a great reduction in computing costs, and more recently the microcomputer, based on LARGE SCALE INTEGRATION (LSI) TECHNOLOGY, is revolutionising the computer world. It is now possible to hold a powerful but inexpensive computer in one's hand.

The combined effects of escalating material, energy, and numerous other general costs which are adversely affecting industry have brought about a much greater emphasis on design. The spectrum of the word 'design' covers a wide range of activities including the initial generation of ideas, the creation of geometric shapes, performance calculation, and the processes of manufacture. Design for manufacture is becoming more important as computers are finding their way onto the shop floor in the shape of COMPUTER NUMERICAL CONTROL (CNC) machine tools. Finally, computers are allowing designers to replace work traditionally performed by experimental development inasmuch as some development activities can now be simulated by using a computer which processes as a mathematical model as an analogue.

1.2 DESIGN

It is first necessary to discuss the process of design before we can discuss how computers can aid the designer. Design in general terms can be defined as the means by which solutions are contrived for the solution of engineering. This definition draws no distinction between the traditionally separate fields of design and manufacture, because the boundary between the two is often vague.

There are three fundamental methods of design, and any one may be used in varying degrees on a particular design process. The three methods are:

a) The iterative (or trial and error) method.
b) The direct method.
c) The design selection method.

a)   The iterative design method

This method is frequently used, especially in the early stages of design. The process consists of employing intuition and experience, from which a set of rules, drawn from experience or direct design, to produce a preliminary solution. The resulting design is then analysed to determine if it meets the specified design constraints. If it fails to do so, then it is modified on the basis of further information obtained from the analysis, and reanalysed until the design can meet the specification, or until the designer is convinced that this design is not feasible within the given constraints.

The iterative process, illustrated in flow chart form in Fig. 1.1, therefore contains three main activities:

 i) Preliminary design.
 ii) Analysis.
iii) Redesign.

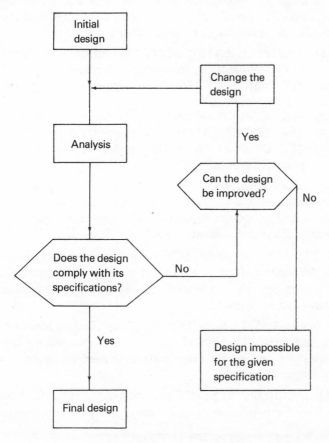

Fig. 1.1 – The iterative design method.

b) The direct design method

The iterative method uses the results of a design analysis as a base for redesign decisions. The direct design method uses the design analysis as a method for effecting the design; in other words, a design is produced directly by analysis. This process implies that a design can be defined by a parametric representation, and that interactions between parameters are known.

c) The design selection method

A design process may yield a number of solutions which all satisfy given constraints. This is applicable to the situation where the design constraints themselves
contain one or more variables. In this situation the selection of the final design,
given the performance characteristics for each design, is largely a matter for
human judgement by compromising performance as it is affected by design
constraints.

The three design methods identify techniques for producing a solution, given
initial constraints. The overall design process goes still further than these three
design methods, having five stages of development:

Stage (1) THE SPECIFICATION STAGE. The definition of design function
and performance constraints or specification.

Stage (2) THE DESIGN STRATEGY. Development of a strategy for arriving
at a final solution.

Stage (3) THE DESIGN SOLUTION.

Stage (4) THE CHECKING STAGE. Checking of the design.

Stage (5) THE APPLICATION STAGE. Application of the design results to
produce a product.

Stage (1), normally called the SPECIFICATION STAGE, consists of formulating requirements for the design with respect to the environment in which the
final product is to be manufactured and used. This stage may require scientific or
intuitive analysis of the environment of the design.

Stage (2) THE DESIGN STRATEGY is based on experience combined with
an approximate analysis of the possible types of design. Decisions taken at this
stage are those of principle: for example, is the machine to be powered by an
electric motor or petrol engine?

Stage (3) THE DESIGN SOLUTION itself uses combinations of the three
basic design methods which we have already discussed. This stage is often divided
into sub-stages, because its use often creates the need for deployment into
component design processes.

Stage (4) The CHECKING STAGE tests the validity of the final design and
ensures that assumptions taken in each stage of the design are still valid at its
completion.

Stage (5) THE APPLICATION STAGE consists of manufacturing the design
or of providing sufficient information to enable the design to be manufactured
without further design effort. This normally consists of providing production
drawings and instructions or, in a highly automated process, a set of machine
settings and operating instructions. The application stage may also include the
testing and installation of the design.

The stages in the production of a design, combined with the activities or
decisions present in each process, can be summarised as follows:

1) *Specification*

   Production of the design objective by the analysis of the environment of the design.

   Formulation of design performance requirements from the analysis of the environment of that design, and from experience.

2) *Design strategy*

   Decision as to the principle on which the design is to be based learned from estimates of the probable performance of various types of designs and from experience.

3) *Design solution*

   Formulation of specifications of component designs.

   Resolving of a final design by the following methods:

   a) Iterative design – by analysis and readings.
   b) Direct design – by an analysis of design parameter interactions.
   c) Design analysis – the selection of a solution from many different possible designs by judgement.

4) *Design checking*

   Verification of results by analysis.

5) *Application*

   Preparation of production instructions. Preparation of testing instructions.

We shall assume that a component design process has no effect on the design of the system of which the component is a part (an assumption which obviously is not always valid). No attempt is made at this point to break down the design solution stage into sub-stages, because this breakdown depends on the nature and complexity of the product being designed.

1.3 COMPUTER AIDED DESIGN

Computer aided design (CAD) is a technique in which man and machine are blended into a problem-solving team, intimately coupling the best characteristics of each. The result of this combination works better than either man or machine would work alone, and by using a multi-discipline approach it offers the advantage of integrated team-work.

CAD implies by definition that the computer is not used when the designer is most effective, and vice versa. This being so, it is therefore useful to examine some individual characteristics of man and computer in order to identify which processes can best be separately performed by each, and where one can aid the other. Table 1.1 compares the capabilities of man and computer for a range of tasks. It can be seen that in most cases the two are complementary, that for some tasks man is far superior to the computer, and that in others the computer excels It is, therefore, the marriage of the characteristics of each which is so important in CAD. These characteristics affect the design of a CAD system in the following areas:

a) DESIGN CONSTRUCTION LOGIC — the method of constructing the design.
b) INFORMATION HANDLING — the storing and communication of design information.
c) MODIFICATION — the handling of errors and design changes.
d) ANALYSIS — the examination of the design and factors influencing it.

TABLE 1.1 CHARACTERISTICS OF MAN AND COMPUTER

		Man	*Computer*
1.	Method of logic and reasoning	Intuitive by exper-ience, imagination, and judgement	Systematic and stylised
2.	Level of intelligence	Learns rapidly but sequential. Unreliable intelligence	Little learning capability but reliable level of intelligence
3.	Method of information input	Large amounts of input at one time by sight or hearing	Sequential stylised input
4.	Method of information input	Slow sequential output by speech or manual actions	Rapid stylised sequential output by the equivalent of manual actions
5.	Organisation of information	Informal and intuitive	Formal and detailed
6.	Effort involved in organising information	Small	Large
7.	Storage of detailed information	Small capacity, highly time dependent	Large capacity, time independent
8.	Tolerance for repetitious and mundane work	Poor	Excellent

Table 1.1 (*C'td.*)

		Man	*Computer*
9.	Ability to extract significant information	Good	Poor
10.	Production of errors	Frequent	Rare
11.	Tolerance for erroneous information	Good intuitive correction of errors	Highly intolerant
12.	Method of error detection	Intuitive	Systematic
13.	Method of editing information	Easy and instantaneous	Difficult and involved
14.	Analysis capabilities	Good intuitive analysis, poor numerical analysis ability	No intuitive analysis, good numerical analysis ability

Let us now consider these four areas which are of such significance in CAD.

a) Design construction logic
The use of experience combined with judgement is a necessary ingredient of the design process. The design construction must therefore be controlled by the designer. This means that the designer must have the flexibility to work on various parts of the design at any time and in any sequence, and be able to follow his own intuitive design logic rather than a stylised computer logic.

The computer cannot cope with any significant learning. This must be left to the designer, who can learn from past designs. The computer can, however, provide rapid recall of old designs for reference. Thus, in some ways the designer can pass on his experience to the computer, and other designers can then have access to it.

b) Information handling
Information is required from the specification before the design solution stage can proceed. Similarly, when the design solution is complete, information must in

turn be output to enable the design to be manufactured. Fig. 1.2 shows the application of this process to manual design. Information is assimilated by the designer from the input specification. The design solution process then takes place, whereby information is passed from the designer to paper and back again in the form of sketches and calculations. When this process is completed, manufacturing information in the form of drawings and instructions is produced.

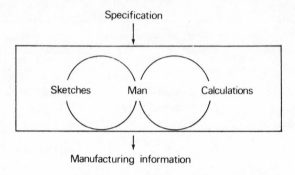

Fig. 1.2 – The conventional design process.

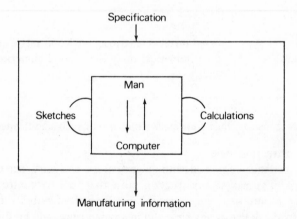

Fig. 1.3 – The design process using CAD techniques

Fig. 1.3 shows the process extended to the combination of designer and computer. The design solution stage now includes a flow of information between the designer and computer in the form of graphics and alpha-numeric characters.

The initial specification must be input to the designer in order that selected parts can be communicated to the computer in a form it can 'understand' and use. The first role of the computer is to check the information for human errors, which must then be corrected by the intervention of the designer.

The human brain is able to store information in an intuitively ordered manner, but its storage capacity is limited, and the information is not all retained as time passes. By contrast, computers have no ability to organise data intuitively but have large permanent storage capacities. Information storage should therefore be carried out by the computer under the direction of the designer.

The output of manufacturing information, from the solution stage of the design process, usually involves the production of drawings. This is a slow and mundane process when carried out manually but is quite suitable for execution by the computer. It is therefore desirable to allow the computer to generate as much production information as possible, so freeing the designer from repetitious work at all stages of the design process.

c) Modification

Design descriptive information must frequently be modified to make correction of errors, to make design changes, and to produce new designs from previous ones. The computer has the ability to detect those design errors which are systematically definable; whereas man can exercise an intuitive approach to error detection. For example, the computer can calculate the torque capacity of a shaft, whilst a designer can tell from experience and judgement that the shaft is too small.

The automatic correction of errors is generally difficult for a computer. It should therefore be left to the designer to monitor corrections of errors and any other design changes which may be needed.

d) Analysis

A computer is very good at performing those analytical calculations of a 'numerical analysis nature' which man finds time-consuming and tedious. As much as possible of the numerical analysis involved in the design should be done by the computer, leaving the designer free to make decisions based on the results of this and his own intuitive analysis.

It can be seen from the discussion so far that there exists a clear division between the functions of man and computer in CAD.

The computer has three main functions:
1) To serve as an extension to the memory of the designer.
2) To enhance the analytical and logical power of the designer.
3) To relieve the designer from routine repetitious tasks.

The designer is left to perform the following activities:
1) Control of the design process in information distribution.
2) Application of creativity, ingenuity, and experience.
3) Organisation of design information.

REFERENCES

[1] Sutherland, I. I. SKETCHPAD: a man-machine graphical communication system. *Proc. SJCC* 1963, 329, Spartan Books, Baltimore, Md.

CHAPTER 2

The Digital Computer as a Design Aid

2.1 THE COMPUTER

Computers are now in common use both in scientific and commercial fields. New electronic components such as integrated circuits have led to the development of computers that have considerable computing power and which are physically small, reliable, and sufficiently low in cost to make their use acceptable for many applications.

Computers are in themselves systems, consisting of a **Central Processing Unit** (CPU), surrounded by various devices called **Peripherals.** The computer system is completed by computer programs which allow it to operate and perform calculations. The physical component units making up the computer are often referred to as **hardware,** and the computer programs **software.**

The CPU is the heart of the computer. It contains three main sections; the **controller** for sequentially examining each instruction and directing the action of the computer, the **arithmetic unit** for performing additions and subtractions which are fundamental to computers, and **core store** where programs and data are stored for immediate use. Core storage has, in the past, been expensive, so that peripheral devices are used for mass storage.

The peripheral storage device, sometimes called the **backing store,** may be of several types, with costs being dependent on capacity and the speed of access by the processor. The most common high speed mass storage device is the magnetic disc, usually of the cartridge type which so permits easy disc interchange. The disc may be used for storage of programs, data, or both together. Access time on modern discs is sufficiently fast to permit the core storage to be extended in an 'imaginary' fashion out onto the disc, creating a system known as **vitual memory.** Generally, disc access times can be measured in a few microseconds.

Disc capacity is normally measured in terms of words or bytes of information that can be stored.

It should be noted that for most computers, the **byte** is defined as: 1 byte = 1 alphabetic character = 2 digits = 8 bits, and that some computers have fixed word lengths of 16 bits or 32 bits, while other computers have variable length words in multiples of 1 byte or 8 bits.

The capacities of disc units vary from unit to unit, typical storage capacities ranging from 1 million up to 200 million characters per complete disc unit.

The big advantage of disc storage over most other mass storage devices is that information can be randomly accessed and transferred into core at typical rates of 1 million characters per second. Thus the magnetic disc is used for storing sub-programmes which may be rapidly swapped in and out of core as required, or for data such as library information which is frequently required.

Inexpensive disc units having capacities of 300 K bytes are available, and they utilise a relatively soft plastic material for a disc, as opposed to the conventional disc which resembles a gramophone record. This inexpensive device is known as a **floppy disc** and is now associated with the new microcomputer systems.

A second common mass storage device is the magnetic tape unit. Information such as programs or data may be written onto tape at speeds in excess of 300 000 characters per second. This information can be transferred back into core at the same rate. However, the problem with magnetic tape is that it can take a significant time to find the required information before transference can take place, because the tape must be searched sequentially. Thus access times can occupy many seconds, which makes magnetic tape unsuitable for library work where rapid data retrieval is a necessity.

Magnetic tape is best used for dumping information for permanent filing, for transferring information to another computer or machine, and for storing information where it is required in a sequential fashion. It is somewhat less expensive than other forms of storage, and its storage capacity is normally very large; for example, one reel of tape may be capable of storing over 1.5 million characters.

Increasing use is now made of the cassette or cartridge magnetic tape system in computer systems. While these have much less storage capability than the conventional IBM industrial compatible magnetic tape units, they do have the advantage of being inexpensive and convenient for certain applications; for example, programs or data can conveniently be stored on a cassette so that each user of the computer can keep his own files.

The least sophisticated forms of storage device are punched paper tape and punched cards. Information can be punched onto paper tape or cards at typical rates of 100 characters per second and read at faster rates. Punched cards are normally used in program development; changes can easily be made in the program by inserting a card containing the appropriate instruction. Punched paper tape is most commonly used as a medium for transferring data or program information to a numerically controlled machine.

Finally, increasing use is being made of the **teletype** or alpha-numerical **Visual Display Unit** (VDU) with keyboard as an input/output device to a computer. Such devices are connected directly with the computer so that the user can communicate with the computer by typing instructions on the keyboard.

The computer can answer by printing information or displaying it on the VDU. Many computers operate with a number of teletypes or VDUs with the computer operating in a **time-sharing** or **multi-access** mode so that more than one user can use the computer at the same time. In this type of system the computer is sufficiently fast to permit the users to work simultaneously without noticing any effect of other user's work. Where large amounts of alpha-numeric information are required to be output from the computer, then a **line printer** is used. A line printer can print up to 137 characters per line, and some can print at rates of 1000 lines per minute.

The computer's capacity to solve a problem requires that the problem must be thoroughly defined, so that the exact sequence of operations for obtaining a solution can be written in a form which the computer can inderstand. The planning of such a sequence is called **programming**, and the actual plan is known as a **program**.†

Programming in terms of adding and subtracting would be very tedious and, since many of its functions such as multiplying and dividing are commonly used these functions are specified by codes which form a **programming language.** Each language requires a **program language compiler** which is fed to the CPU before any program can use the language, and each computer also requires a system program which relates all the peripherals to the CPU.

Once a computer has a system program and language compiler, any person can use the computer provided that he learns the language. One of the most common language isFORTRAN, which can be learned in a very short time.

2.2 USING THE COMPUTER AS A DESIGNER'S AID

The conventional tools of the designer have in the past been limited to the drawing board or draughting machine, drawing instruments, and the slide rule. The designer is concerned with information from design data sheets, drawings, parts lists, and technical references, and with solving equations or formulae. Furthermore, to all this quest for information must be added his experience and knowledge.

In recent years the designer's slide rule has been supplanted by the pocket calculator, and many designers have in addition had access to a computer, uaually via a teletype terminal.

We shall now discuss the designer's functions in relation to the computer in order to see where the computer may be used effectively to aid the design process. The principal areas of application are as a calculator, a data bank, and a draughting aid. We shall first discuss each application individually, and then the relationship between the three applications via the computer.

---

† Spelt thus, not 'programme'.

2.2.1 Using the computer as a calculator

The designer has traditionally used data from technical data sheets which are found in the codes of practice issued by national organisations, such as the British Standards Institution or the American Society for Mechanical Engineers. Some of the information contained in these data sheets has been built up from practical experience and from experiments linked by empirical relationships. Very often the use of these data sheets is time-consuming and costly. Furthermore, calculations by designers are often subject to human error even when the calculations are relatively simple. Some drawing offices have found it worth while implementing some often-used codes of practice on a computer. Where codes have been programmed, the designer is often not faced with the task of interpolating from graphical or tabular data, but can calculate results for himself. Experience has shown that where computers have been used, the results have proved to be more accurate and reliable.

There are situations where calculations exceed the capabilities of hand calculations with a pocket calculator, owing to the complexity of the problem, or because many different alternatives have to be computed carefully and the results compared to determine an optimum. If existing programs are available, the designer only has to justify the economic cost of using the computer for his problem. In addition, if no program exists, he may have to decide whether he or someone else should write a program. Because programming is expensive, such decisions cannot be taken lightly.

The computer can be used as a calculator at two stages in the design. The first stage is the evaluation of the information relating to the **specification,** and the second is the calculation of values during **synthesis.** The specification stage will deal with quantities such as load and speed for the transmission of a hydraulic motor, input hydraulic oil pressure and flow to achieve the specified power output. The synthesis stage involves the engineering of the design as far as the preparation of working drawings, schedules, and any other information needed for the manufacture of the design. It is usually in the synthesis stage that the computer can be of most use. At this point the design must be assessed for optimum performance and minimum costs. This usually entails the provision of many sets of calculations in order to iterate to the best possible design. The accuracy of the computer will permit the results of minute variations of a design to be compared, with the result that judgements can be then made on reliable quantitative data.

A good example of using a computer in a design synthesis is the application of finite element stress techniques to complex pressure vessel design or car body design.

One of the main reasons why digital computers are not used by many designers is the difficulty in describing the problem to the computer; that is, the data preparation can be time-consuming and difficult. We shall see later that

CAD techniques which combine graphics and calculations, go a long way to overcoming problems of data preparation.

2.2.2 The computer as a data bank

We have seen in Chapter 1 that the computer has a reliable memory. Utilisation of this virtue, combined with suitable mass storage devices, makes the computer a useful tool as a data bank. Many companies now store details of products on computer files. We shall see later that, when the computer is coupled to a graphic display, the graphical data can be stored along with alpha-numeric data to provide a very powerful data bank system.

The success of data banks lies in their method of structuring the data and in the speed at which information can be retrieved. There are now many methods of storing data for fast retreival. A tree-like structure is popular: it contains pointers or by-words so that a user can quickly trace a path through the data bank to the required information. A graphics display can be used to great advantage because large amounts of information can be quickly displayed, so making use of mass visual sensing to assist in decision making.

2.2.3 Using the computer as a draughting aid

An important part of engineering is the formation of graphical data. Drawings have always been the main method of communication in the engineering world. although in some specialised areas where numerically controlled machines are employed, this method of communication may change.

Designers have traditionally developed their final designs via rough preliminary sketches. It is in turning the rough sketches into working drawings that the computer can play a useful role, because this is a tedious and time-consuming process. Furthermore, a large part of draughting is concerned with the use and modification of existing designs. Once graphical data has been stored in the computer it can be re-used, modified, and edited, and parts can be added to produce a new drawing. The computer will play a significant role in this area.

2.3 EXISTING CAD TECHNIQUES

It is now important to consider how the computer might be used as a CAD tool to provide a link between draughting, data banks, and calculation. The method of using the computer or processing information, and the employment of devices and techniques which the designer might use to communicate with the computer, will play important roles here.

2.3.1 Processing techniques

The computer can be used in a number of ways to aid in almost all stages of the design process. A number of techniques are available for communication between man and machine, each of which has different characteristics (see Fig. 2.1):

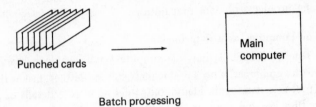

Punched cards

Batch processing

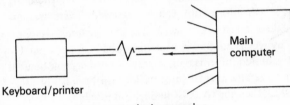

Keyboard/printer

Remote terminal processing

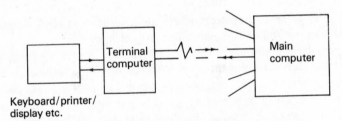

Keyboard/printer/
display etc.

Intelligent terminal processing

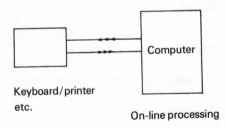

Keyboard/printer
etc.

On-line processing

Fig. 2.1 – Processing techniques.

a) **Batch mode processing**
This method of computing is usually used for straight forward computing processes involving no interaction between the user and the machine.

The user normally prepares data and machine operating instructions on punched cards, reads them into the machine and, sometime later, receives a print-out of results which he can examine at leisure. The results may also be received as a graph or drawing prepared by the computer's graph plotter.

b) **Remote terminal processing**
A degree of interaction between the user and the computer can be introduced by using the computer from a remote terminal via a telephone line. The terminal may comprise a simple keyboard and printer, or a more complex keyboard with a graphics display and Xerox-type hard copy unit.

The remote terminal is normally used for running programs which require and produce only small amounts of data, but which need to interact with the user.

The user can call up his program from the terminal and run it. He can then supply it with data at its request and have results transmitted back to him almost immediately.

c) **'Intelligent' terminal processing**
The 'intelligent' terminal normally consists of a small computer with one or more peripheral devices, connected via a high-speed line to the large central computer.

The use of this type of terminal is much the same as for the remote terminal, except that a wider range of communication devices can be used and a larger amount of data can be handled.

The terminal computer could have connected to it a refresh display (see Section 2.3.2) with a light pen, card readers, magnetic tape etc., and could handle many users simultaneously.

d) **On-line processing**
In an on-line mode of operation, the user communicates directly with the computer through its peripheral devices, and he can use all the available devices to their best advantage.

On-line processing is normally used where a high degree of interaction is necessary.

2.3.2 Graphical input and output devices
There are many design processes that use graphics. A brief description is now given of some of the most common input and output devices.

a) Refresh display and light pen
Many CAD systems are based on the use of the **refresh graphics display**. This

consists of a cathode-ray tube (CRT) which gives a line diagram on a screen that is similar to that of a television receiver. However, a television picture is produced by scanning the entire screen with an electronic beam. The intensity of the beam is modulated during scanning, and the picture is then made up of a series of parallel lines. This type of display is known as a **raster** system. But most graphic displays used with a computer are concerned only with discrete lines, therefore it is no longer necessary to scan the whole screen. The beam merely traces out each line on the screen. Thus, to position a spot on the CRT, horizontal and vertical (x and y) deflection voltages are applied to the beam in proportion to the deflection. The brightness is controlled by an externally supplied voltage (z-modulation). In most systems only two levels of brightness are used, bright-up or zero.

The electron beam striking the rear of the screen causes the phosphor coating at that point to glow when the bright-up voltage is applied. This glow has a short but finite duration. For thr picture to be maintained, it must be retraced before the eye can detect any decay in the glow. If the retracing frequency falls substantially below 50 Hz, then the picture will flicker, causing strain to the viewer. Thus, with the refreshed display, the computer must repeat the picture many times a second, and if the picture is complex or the computer has also to perform many additional calculations, flicker could result. To overcome this a small computer is often used in addition to the main computer; the small computer is used only to refresh the picture. The computer performing the refresh task is normally called a **display processor**. A typical system is shown in Fig. 2.2.

When straight lines or characters are to be displayed, special hardware facilities are often utilised in the refreshed display system to provide a fast response. The straight line or vector is produced by generating sucessive spots along the line, this process being repeated to give a continuous image. Characters may be generated by software subroutines, but a hardware character generator is often used to speed up the process. Thus, a whole picture is made up of spots, lines, and characters in a suitable order.

The data describing the image to be displayed is usually stored in a display file. While the computer is running the program for the relevant application (user program or applications program), it makes additions or subtractions to or from the contents of the display file. The display processor reads the data from the display file and executes the operations needed to display the corresponding picture, and also refreshes the image as often as required. The display file is a structure of data which can be continuously changed by the user program specifying the image; the display processor is a set of actions which can be program controlled.

Interaction with the display to modify a picture is achieved by the use of a light pen, a cursor driven by a joystick, a keyboard, or combinations of them. The system combines the manipulation of a tracking cross with a light pen. The cross or cursor can be moved to any position on the screen by the tip of the

Fig. 2.2 — A DEC graphic-II GT46 Stand-Alone graphics system.

light pen, and it provides a pointer for indicating the part of the picture which is to be modified.

b) Direct view storage tube (DVST)

The **direct view storage tube** is one solution to the problem of the CRT's image, and consequently to the problem of having to refresh the display. The DVST is similar in construction to the conventional electrostatically controlled CRT, but with an additional grid electrode and a second electron gun called a 'flood' gun. The most common DVST is manufactured by Tektronix, who have deposited a grid electrode as an integral part of the phosphor layer on the screen, giving a bistable phosphor.

The operation of the storage tube depends on secondary emission due to electron bombardment. The grid electrode is negatively charged by the application of a suitable voltage before writing. When the main write beam of electrons is moved so as to display a line, the grid becomes positively charged in the area where the tightly focused beam strikes it, so allowing electrons from the flood gun to be accelerated through onto the screen, causing the phosphor to glow. The image is displayed continuously without the need for any retracing of the write beam, thus making little demand on the computer. The image is erased from the screen by recharging the grid negatively. Selective erasure is difficult with the DVST, and limited movement effects can be displayed only by rapid replacement of the image.

It is possible to operate the Tektronix storage tube in a write-through mode so that a cursor may be displayed without storing it. The cursor may be driven by a joystick system or by thumbwheels at a keyboard console. A Tektronix type 4014 display is shown in Fig. 2.3. It is often used in conjunction with a Xerox type hard copy unit so that a picture displayed on the screen can be recorded in seconds.

c) Digitiser

A common way to take x, y coordinates off a drawing is to use a digitiser. The digitiser consists of a table or board similar to a drawing board, and a probe consisting of a pen or cursor which can be moved over the surface. The pen, or cursor, contains a switch that enables the user to register x, y coordinates at any desired position. These coordinates can be fed directly to a computer or to some off-line storage device such as industrial compatible magnetic tape.

There are many digitisers on the market, operating on various principles. The Talos digitiser, for example, shown in Fig. 2.4, is based on an accurately set out wire grid mounted on a temperature-stable board. An electric field is electronically switched to portions of the draughting surface in accordance with the movement of the pen or cursor — which, in turn, senses the electric field. Utilising a writing servo principle that can, in fact, resolve much larger areas of active surface, the digitiser surface is driven in increments of only one inch by one inch.

When the pen or cursor goes beyond the dynamic range of the one-inch area, the activated area is electronically switched to follow the pen or cursor. The activated area is resolved by circuitry so that interpolation takes place in the x and y directions. This technique does not place any restrictions on the size of the digitiser, since the system resolution is independent of size.

The system has the advantage that any specific position is defined by a data output that represents the absolute location. There is no need to reference the cursor or pen to a starting position, thus it can be removed from the surface and replaced, with an immediate output of the absolute position. The Talos digitiser is abailable in sizes from 11 x 11 inches up to 44 x 60 inches, and an accuracy of ±0.005 inch is possible.

Fig. 2.3 – Tektronix 4014 Direct View Storage Tube Display. Display sizes 15 in (38 cm) by 11 in (28 cm).

Fig. 2.4 — Talos digitiser with multi-button cursor system (Courtesy of Sintrom Electronics Ltd.).

d) Tablet

The tablet is a small low-resolution digitiser. It is often used as an alternative to the light pen or in association with storage tube displays. The tablet served as a surface corresponding to the CRT but remote from it. The user writes on the surface of the tablet, and the position of the pen is transmitted to the computer. The screen which then displays a cursor is an echo of the position of the pen on the tablet.

The tablet is particularly suitable for interactive design since it allows the designer to work naturally with a pen in his hand while watching the screen to check his actions. The computer can straighten lines and present a clean picture if it is fed with accurate salient points. These points can be either input from an alpha-numeric keyboard or from a menu on the tablet. A menu consists of defined squares on the tablet. Any point digitised within a square activates an appropriate instruction within the computer.

The digitiser is much used in draughting applications where information is taken off accurate drawings. It can be useful in working on large layouts. However, tablets are used more and more where increases in computing power, coupled with large and faster display systems, make of the digitiser less essential.

A well known tablet, on which many others have been based, is the Rand design. This has a wire grid of 1024 closely spaced wires in both x and y directions. The wire grid is mounted in a square plate. The wires are pressed in touch with a set of contact pads which are laid out in the margin of the work area and which provide a voltage drive. The pads are arranged in the Gray-code pattern for 10 digits, so that by energising each of the 10 sets of pads in turn, each wire receives a unique pattern of pulses. This contact configuration is applied in each coordinate direction. The voltage pulses are sensed by a capacitive probe which reads the pattern of pulses from the wire nearest to it. The Gray-code signals received by the probe are decoded into binary values, and by subtracting each reading from the one before, the computer can check for errors or pen lift-off, since each signal should differ from the previous one by 1 bit only.

The wires in the grid must be spaced at intervals of 0.01 inches in a typical 10-inch square tablet. Furthermore, each wire must be accurately positioned, and this can lead to an expensive tablet.

The Talos digitiser in low resolution form (±0.01 inch) with an 11-inch square working area, makes an ideal tablet combining low cost with an excellent all-round performance.

There are many other types of tablet such as the sensitive film design developed by Walker at Reading University and the Bell Telephone Laboratories. Another tablet is the Graf/Pen Sonic Digitiser which is based on two linear microphones which sense acoustic signals from a sonic source located in the digitising pen or cursor.

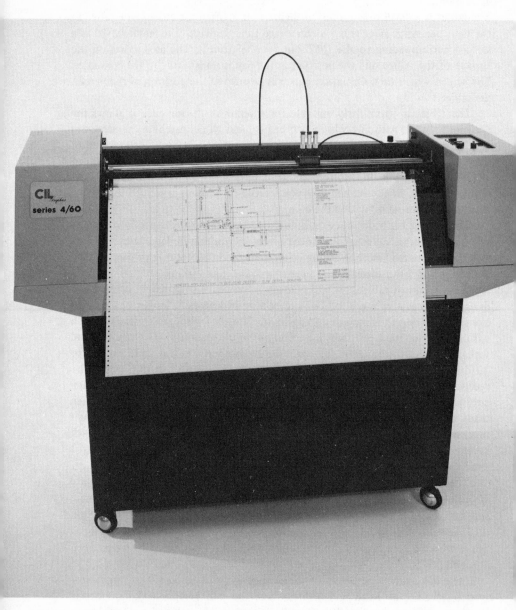

Fig. 2.5 – A typical drum plotter with a paper width of 36 inches, 0.025/0.05 mm increment size, and 15 to 30 cm/sec axial drawing speed.

e) Plotters

Hard-copy output from a graphics system is generally produced by a plotter. Plotters come in various forms ranging from the drum plotter to the flat bed and the electrostatic plotter.

One of the least expensive, and consequently the most common, is the drum plotter. It consists of a drum driven by a stepper motor, with the paper fed over the drum. The paper has perforations along each edge which engage in sprockets at either end of the drum. A pen holder, containing one or more pens, is mounted on a gantry above the length of the drum. A second stepper motor is used to drive the pen unit along the gantry, and the pen is moved up or down by a solenoid. A typical drum plotter with a paper width of 36 inches is shown is Fig. 2.5. The plotter can be driven from a magnetic tape unit (off-line) or directly from a computer (on-line).

A more expensive type is the flatbed plotter. The gantry is supported over a flat surface, and with a second system of bearings, motion in two dimensions is produced. The paper is laid flat on the bed, and is usually held down by vacuum or electrostatic means. The design can be followed as it is being drawn. Again, the plotter can be driven in an off-line or on-line mode.

A typical flatbed machine is shown in Fig. 2.6. In this model the gantry and pen unit are driven by d.c. servo-motors with optical encoders for positional sensing. A Texas microprocessor is used as a control unit and is built into the plotter. The microprocessor allows certain shapes, such as circles and ellipses, to be preprogrammed, so that simple instructions to the plotter will quickly produce a complicated shape. The microprocessor is therefore used as a vector generator, and it has the advantage that variations in standard shapes can be programmed into the microprocessor to suit a variety of applications.

There are now a number of small flatbed plotters, such as the Tektronix 4662 or Hewkett - Packard 7221A and 9872A. These plotters open up a new field of interactive computer graphics because of their low cost. They are capable of being used over a telephone line for on-line graphical communication. Many of the small flatbed plotters incorporate a microprocessor which permits relatively simple instructions to be transformed into a variety of graphical shapes.

Finally, the third common plotter is the electrostatic plotter shown in Fig. 2.7, consisting of an electronic matrix which can print dots onto charge-sensitive paper. The matrix is held on a bar, and the paper is fed over rollers and under the matrix. The drawing is produced in a raster type format, as the paper passes under the matrix. The dots can be produced with a density sufficiently high (200 dots per inch) for a series of dots to make up a line. The plotting speeds are very fast, and plotting widths up to 72 inches are possible. However, the electrostatic plotter requires special paper and chemicals. It is also necessary to preprocess the drawing data so that it is in a raster format. The electrostatic plotter can also be used as a printer.

Fig. 2.6 – A typical Ao flatbed plotter with a drawing speed up to 30 cm/sec, 1 g acceleration, and in-built level 2 micro-intelligence. (Courtesy of Computer Instrumentation Ltd.)

Fig. 2.7 – A typical electrostatic plotter with 11-inch wide paper, interfaced to a minicomputer. (Courtesy of Calcomp Ltd.)

CHAPTER 3

Interactive CAD Systems

In the previous chapter we discussed devices which are used for input and output of graphical and design information to and from a computer. In this chapter we shall discuss how individual peripherals may be linked to a computer to form a system which can be used for computer aided design. Two systems will be described, one based on a refreshed graphics display and another based on a storage tube display and digitiser.

3.1 A REFRESHED GRAPHICS DISPLAY SYSTEM

Digital Equipment Corporation, Maynard, Mass. USA, have produced a series of low-cost refreshed graphics display systems that started with the GT40 Graphics Terminal and continued through the series GT44, GT43, and GT46. The GT40 type systems are based on the use of minicomputers, and they are particularly aimed at a mass market since their prices are lower than those of many refreshed graphics systems.

The latest version of the GT40 series, the GT47, consists of the following units: a PDP 11/34 central processor unit containing CPU and store, a VT/11 display processor unit, a keyboard unit, a CRT display unit, and a light pen. Various mass storage units such as discs and magnetic tape units complete the system. The system can also be interfaced to a larger or main frame computer (called a **host computer** in this case) via a communications interface. A typical system arrangement is shown in Fig. 3.1.

The PDP 11/34 system has an important feature — a multichannel highway called a **Unibus** which interconnects the various parts of the system; it also allows other parts of the system to effect transfers of information without the intervention of the CPU. These transfers are called **non-processor requests** or NPRs.

The VT/11 display processor is an autonomous unit which controls the display and light pen. When it has received an instruction to start it operates in a similar manner to the CPU, obtaining instructions and data from the memory of the PDP 11/34 using NPRs.

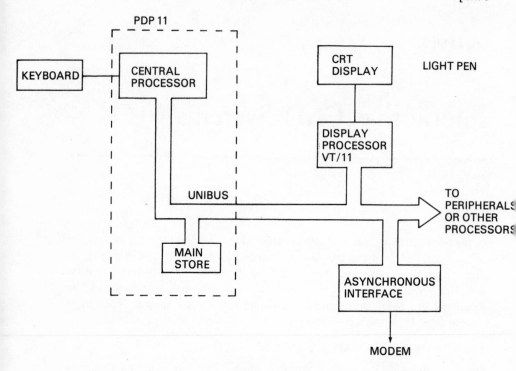

Fig. 3.1 – GT46 interactive graphics display system.

The display file is organised and maintained by the CPU which interprets the graphics instructions originating from the graphics software. The CPU places the display file instructions and data in sequence in its memory store. To start the display, the CPU sends the address of the initial display instruction along the Unibus to the program counter in the display processor. The program counter contains the store address of the next instruction. The display processor performs the program from this first instruction until it is stopped by the CPU.

The display processor contains four main registers: program counter, status register, X position, and Y position. The program counter increments through the range of memory addresses by means of the clocking system of the display processor unit (DPU). It can also be instructed to 'jump' addresses.

The status register performs the same functions as the order register in a conventional computer. It receives instructions and decodes them. The decoding of the status instruction consists of calling for data from the Unibus and allocating it to the appropriate data register.

There are two main sets of instructions: a set for control and a set to organise the display processes. The control instructions are: set graphic mode, jump, no-operation (no-op), load status register, A or B. This set controls the cyclic operation of the DPU. 'No-op' instructions have no effect; they simply take the place of unwanted instructions in the core. Load status instructions control facilities such as the light pen or fixed increments in graph plotting.

The graphic instructions are: vectors, point indication, character generation, and graph plotting. The display graticule resolution is of 10 bits, which gives 1000 points in each direction. Point positioning on the drawing of long vectors requires 10 bits, of a 16 bit word, in both data registers. The remaining 6 bits in each register are used as identifiers and for point or line bright-up.

The display functions in the display processor

There are a number of display functions such as the generation of points, characters, and vectors which are implemented by the DPU. These will be discussed for various applications of the system.

Point mode

The CRT beam is deflected to a point by x- and y- deflection voltages from two summing amplifiers. The inputs to these amplifiers are from various function-generating devices. The most direct function is point mode. The x and y co-ordinates are stored in the data registers called X and Y position-hold registers. The outputs from these registers are fed to the deflection amplifiers via digital to analogue converters. The maximum time of CRT beam travel is 21 μs before the bright-up pulse is given. This method of beam positioning is also used as the initial positioning instruction in vector or character generation.

Graphplot

The graphplot mode works on the principle that one coordinate (the abscissa) advances in equal increments, while the other (the ordinate) has an arbitrary value. Thus one of the load instructions sets the increment for the x and y abscissa when graphplot mode is called. The ordinate values are successively called from the CPU store and displayed. This technique is used for graph plotting because of its speed and simplicity.

Character mode

The data registers are of 16-bit length, which allows the normal form of character output to be in pairs since two 8-bit character bytes can be hald in one data register. The character generator is based on a ROM (read only memory) giving an 8 x 6 dot character matrix in a vertical-scan mode. The scan is in fact bidirectional to give maximum speed, the scan commencing from the bottom left-hand corner, scanning up the first column, downwards for the second, and so on.

The character set consists of 96 ASCII (American Standard Code for Information Inter Change) symbols and 31 special patterns for Greek symbols and mathematical symbols. A number of lower-case characters have been added to the character set. Furthermore, the characters can be displayed in italic format. Both of these features greatly enhance the presentation of alphanumeric information.

Vector modes

Vector generation is important in computer graphics since the speed of presentation of graphical information can be greatly enhanced. The vector drawing techniques in the system make use of a combination of analogue and digital or hybrid devices. There are two vector modes, long and short; they differ in that for short vectors one data word is used to hold both x and y increments, while for long vectors two 10-bit numbers are necessary to describe the vector, requiring the transfer of two data words.

When a vector is drawn, point mode is first used to define the start point; this sets up the x and y coordinates in the holding registers, where they are converted into analogue signals to position the CRT beam. The horizontal and vertical displacements Δx and Δy are transferred from memory into the data registers. The Δx and Δy values determine the gradient of the vector, and they are also used to give the termination point.

Two ramp functions are generated to deflect the beam along the required path by applying a 10-bit binary rate multiplier (BRM) to both Δx and Δy. The outputs from the BRMs are a succession of pulses of frequency proportional to Δx and Δy. The pulses are fed to integrating amplifiers which generate the appropriate voltage ramps. These ramps are added to the initial x and y coordinates, causing the beam to deflect along the required path. The larger of Δx or Δy is also loaded into a counter, the down-count register. Thus the vector process is terminated when the down-count register reaches zero. At this point the X and Y position register contents are placed into the X and Y hold registers so that they may be used as the start point of the next vector.

The pulse-rate output of the BRM is proportional to the magnitude of the number fed into it. If Δx and Δy are small, the output rate will also be small, and the vector will be drawn slowly. To get round this problem, the Δx and Δy register contents are normalised, before the vector-generation process is started, by left-shifting them in their registers until the larger one has a 1 in the most significant digit position. The two registers will then contain the largest values of Δx and Δy which are in the correct ratio to define the vector gradient. This ensures that the vectors are generated at maximum speed regardless of length.

3.2 AN INTERACTIVE CAD SYSTEM BASED ON A STORAGE TUBE DISPLAY AND DIGITISER

A stand alone interactive CAD system was developed at Imperial College in 1969 by Besant and Jebb. It used a digitiser as a graphics input device into a Digital Equipment PDP 8/I computer with a Tektronix storage tube as a visual display unit. The digitiser was based on a D-Mac pencil follower. Operation of the D-Mac is based on an electro-mechanical servomechanism which is located beneath the digitising table top. The mechanism consists of a gantry which can be driven in the X direction by a servo-motor, and a head unit containing three magnetic sensing coils which are mounted on the gantry and are driven by a second servo-

motor in the *Y* direction. A cursor containing a coil (Fig. 3.2) which sets up a
magnetic field, is used by the operator. The coils on the gantry head sense the
position of the cursor and provide a signal which ensures that the gantry head is

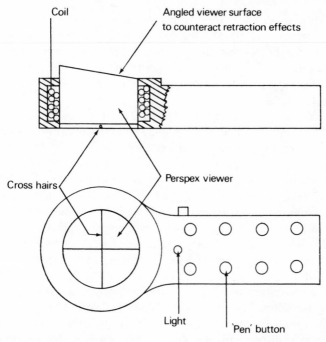

Fig. 3.2 — The CADMAC digitising pen.

driven directly beneath the cursor. Whenever the cursor is moved, the gantry head follows it. Two optical encoders are used to monitor the X and Y positions of the gantry, and pulses from these encoders are continuously monitored by the computer.

The digitising table has been further developed into a combined digitiser-plotter. A glass top was fitted to the digitiser; it can be raised from the front to provide access to the plotting surface which is directly below the gantry unit. A multiple pen unit was attached to the gantry head unit. The gantry system is driven by the servo-motors, under computer control, to give a conventional flatbed plotting system.

A number of commercial systems, based on the use of PDP 8 computers have been sold by Computer Equipment Co Ltd. The company expanded as a systems supplier to become Cetec Systems Ltd, and then Ferranti-Cetec Ltd.

In 1972 the CAD work at Imperial College switched to PDP-11 computers to take advantage of a 16-bit computer as opposed to the PDP-8 which has a word length of 12 bits. Systems have been evolved round a PDP-11/20, an 11/40, and an 11/45 computer. There are many similar systems based on a work station consisting of a digitiser, storage tube, and minicomputer. Many of these systems have originated in the United States as a result of work in the electronics field. Manufacturers such as Computer Vision, Applicon, and Calma are now well known in Europe. Although European systems manufacturers have not been so successful as their American rivals, independent organisations such as the civil engineering consulting engineers, Scott Wilson Kirkpatrick and Partners, have successfully developed their own CAD software to operate on a PDP-11/40 computer with a digitiser and a Tektronix 4001 display. This system has been operating successfully in their drawing office for a number of years and is being used for general draughting and reinforced concrete detailing.

The CADMAC 11 system at Imperial College, shown in Fig. 3.3, will now be described since it is fundamentally similar to many other systems using comparable CAD peripherals. A block diagram of the system hardware is shown in Fig. 3.4.

The computer used is a D.E.C. PDP-11/45 consisting of 24K 16-bit words of core storage, two 1.2 million word moving head disc units, and an industry-compatible magnetic tape unit. Interfacing between the computer and graphics peripherals is performed via a CAMAC system which contains a data highway similar in many respects to the Unibus on the PDP 11. Communication between the UNIBUS and CAMAC dataway is performed by the CAMAC **crate controller.** CAMAC was used as the interfacing system because much of the hardware such as vector and character generators were in existence as standard modules before such facilities became readily available for graphics applications. Present graphics peripherals can now be easily interfaced directly to minicomputers. For example, the Tektronix 4014 storage tube display contains vector and character generating hardware enabling it to be driven directly from a minicomputer such as a PDP-11.

Fig. 3.3 — A typical CADMAC interactive CAD system.

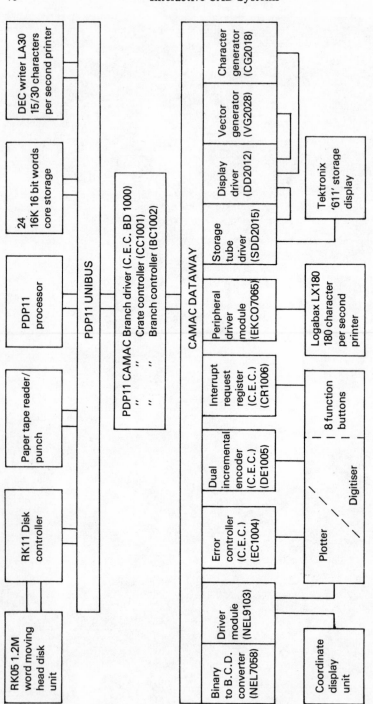

Fig. 3.4 – Block design of the CADMAC 11 system hardware.

The functions of some of the CAMAC modules are now briefly described, since these functions are important whether they are in CAMAC or any other hardware format.

The CADMAC display system

The CADMAC storage tube display is produced by a set of four CAMAC modules:

a) Storage tube display driver (S.E.N. SDD2015)
b) Display drive (S.E.N. FDD2012)
c) Vector generator (S.E.N. VG2028)
d) Character generator (S.E.N. CG2018)

The storage tube driver produces control signals for the storage tube which are used to perform screen erases and to switch between store and non-store, and between write-through and non-write-through modes. It also generates delay signals informing the computer when the control function has been completed.

The display driver contains a position register, the contents of which are converted by a digital to analogue convertor into voltages used to drive the display beam. The register can be incremented or decremented by pulses fed to it from an external source, and its contents can be set by the computer. An intensity pulse is sent by the display driver at every change in the contents of the position register, thereby producing a spot on the display.

The vector generator produces increment or decrement pulses for the display driver at a rate high enough to produce a straight line. The direction and extent of the line produced is controlled by (x, y) values sent to the module from the computer (see Fig. 3.5).

The character generator produces analogue voltage levels representing a character from the ASCII character code fed to it by the computer. The voltages produced represents points on a 7 by 7 matrix. The character is produced by sending intensity pulses at appropriate times as the matrix is being scanned (see Fig. 3.5). The analogue levels produced by the character generator are added to the levels produced by the display driver to position the character.

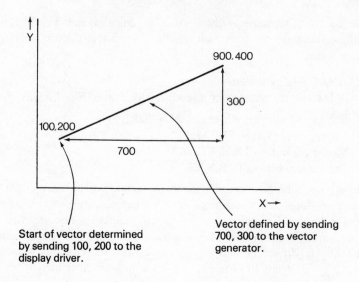

Start of vector determined
by sending 100, 200 to the
display driver.

Vector defined by sending
700, 300 to the vector
generator.

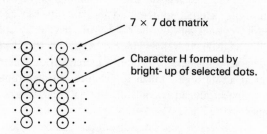

7 × 7 dot matrix

Character H formed by
bright- up of selected dots.

Fig. 3.5 – Vector generation.

CHAPTER 4

A Computer-Aided Design System: The Software

We have seen from the previous chapter that there are two basic types of hardware configurations for CAD. In this chapter we shall discuss in detail how the hardware in a CAD system can be brought together by the use of appropriate software. The philosophy and software described are based on the use of the storage tube, digitiser, and minicomputer, and have been developed over the past seven years by the CAD Unit in the Department of Mechanical Engineering at Imperial College.

Experience in the early phases of CAD development led to the realisation that there was a need for a general CAD system capable of supporting applications oriented to design programs in many different fields.

The concept of a general CAD system was that it should provide:
a)   A system for handling user actions.
b)   A system for the operation of applications programming.
c)   A set of basic system functions and utilities.

It is essential that applications systems can be built into the CAD system without the applications programmer having to be concerned with low-level data, detail systems programming, or pheripheral handling. So far as possible, all applications systems built on the general system should be modular and should conform to a well-defined specification, enabling useful modules in one application to be used in another.

Three classes of system users were defined:
1)   System operators such as draughtsman and designers.
2)   Applications programmers.
3)   System programmers.

System operators have access only to information that allows them to operate the general system and any applications systems. Operation of the general system requires little knowledge of computers or programming.

The applications programmer is allowed information concerning the linking of programs to the operating system and the function of system utilities, but is not allowed enough information to enable him to modify the system. The number of people in the latter class should be strictly limited to avoid systems being modified and made incompatible.

In the development of a general system the requirements of a wide range of applications had to be considered. The design decisions had to ensure that these requirements could be fulfilled by applications programming.

4.1 THE REQUIREMENTS FOR A GENERAL CAD SYSTEM

The general CAD system was developed by considering a wide range of possible uses of such a system. The following were considered in detail:

1) Mechanical engineering design.
2) Building design.
3) Structural engineering design.
4) Electronic circuit design.
5) Animation and graphic design.

It was postulated that four basic processes involving graphics occurred, to various degrees, in each field, namely:

a) Pure analysis – standard design and analysis processes.
b) Pure draughting – production of a drawing or picture by the manual creation and manipulation of lines, arcs, etc.
c) Drawing by analysis – the production of a picture or part of a picture directly from analysis: for example, production of cam profiles.
d) Analysis of drawings – evaluation of the properties of an item described graphically, for example the production of a quantity list by analysis of a builder's plan drawing.

For the system to be able to support pure analysis it must contain facilities for the running of analysis programs of unlimited length and for the storage and rapid retrieval of large amounts of data.

It was considered important that the user should be able to communicate directly and graphically with analysis programs. Graphics facilities were provided which were considered to be sufficient for a general design draughting system. However, the range of graphical construction techniques is so large in practice that the system contained only as many facilities as could practically be incorporated in the draughting system, leaving other more specialised techniques to be developed by the applications programmer.

For both the production of drawing items by analysis and the analysis of drawings, it is essential that there is a simple efficient link between data produced by the draughting system and analysis programs. It is also essential that graphic data can be annotated in a way which is recognised by analysis programs but which does not affect the draughting system.

It was thought that for most *practical* applications the general draughting system would be incorporated in a much larger specific applications system. For this reason the draughting system was as simple as possible consistent with reasonable running efficiency, so that it could be incorporated into an applications system with the minimum of effort.

4.2 THE GENERAL CAD SYSTEM (GCADS)

The facilities embodied in the general CAD system are now described. These facilities are aimed at allowing a user to input graphical information into the computer and file it. Initial data entry is made by digitising rough sketches. The system also permits the user to access the data, manipulate it, process it, output it in hard-copy form, or re-file it for permanent storage. The use of the facilities is described in Chapter 5.

4.2.1 Graphic functions

The input of graphics to GCADS takes place in one of a number of 'symbol' modes. The symbol mode used is selectable from a menu command. The mode provided in the basic system include:

> Line mode
> Arc mode
> Circle mode
> Alphanumeric mode
> Dimension mode
> Rectangle mode

Messages displayed on the screen inform the user of the significance of points digitised in each mode; for example, when the user selects 'arc mode' the message START POINT appears on the screen. Digitising the arc's start point causes CENTRE POINT to be displayed, a further point causes END POINT, and digitising the end point will cause an arc to be displayed and the message START POINT to reappear.

Up to 40 symbol modes can be present in an application system. These may include ellipses, polygons, etc.

4.2.2 Utility functions

A number of functions are incorporated in GCADS to make it easier for the user to operate. These include the following:

> Windowing
> Drive mode
> Angular measurement
> Grid round-off

The windowing process allows the user to specify an area of the drawing he is constructing and to expand it to fill the screen. This enables detail work to be carried out on a small area of a drawing at a large scale.

Drive mode enables the user to specify angles and lengths of lines accurately without digitising them.

The display of X, Y coordinate positions is produced on the coordinate display unit.

By specifying a grid factor the user can, in effect, reduce the resolution of the digitising area. Thus specifying a 300 mm grid means that only points which are on a 300 mm grid from the drawing origin can be digitised. The grid factor is in real units; that is, in table coordinates multiplied by the drawing input scale.

4.2.3 Filing facilities

Files are assigned to a number which is digitised from the menu. A file is brought into use by the menu command FILE TO WORK SPACE followed by digitising the required file number. Data is filed away by the command WORK SPACE TO FILE followed by the file number to which the data is to be assigned.

Other facilities exist for transferring files to and from more permanent named file stores.

4.2.4 Editors

A number of powerful editors are incorporated into GCADS. These include:

Point editor	– allows points to be moved independently, providing the ability to 'stretch' lines.
Line editor	– provides an efficient means of locating and deleting lines.
Macro and symbol editors	– enable macros and symbols to be located and deleted. (Macros are standard components or items that are in regular use.)

4.3 SOFTWARE DESIGN AND ORGANISATION

The system described is based on a single-user system, although multiple-user systems are now emerging. The essential features of the software are very similar in single- and multi-user systems.

4.3.1 Organisation of program and data

a) Operating system

The software for the GCADS system was written to operate under DECs PDP 11 DOS. (Disc Operating System) system. This occupies approximately the lowest 4000 words of core and contains many more facilities than are needed by GCADS. A considerable saving in core storage could therefore be made by either rewriting DOS to contain only the facilities needed, or by cutting these facilities from the existing DOS system. This was not done, for two very important reasons:

1) Changing DOS would severely limit the scope of analysis programs which may be added to GCADS in the future.
2) As the price of core storage is decreasing and the cost of software increasing, it becomes cost-ineffective to spend large amounts of programming time endeavouring to economise on storage, especially if this leads to restrictions in writing future programs.

b) Core utilisation

The core storage available in the PDP-11 for program was divided into two areas called the resident and overlay areas. An overlay is a program resident on the disc which can be brought into core for use by the CPU and then returned to disc. This technique permits the use of many or even large programs to be seen in a computer with small core storage capacity.

All user programs and system operations are called as overlays and operate in the overlay area. The resident area is used to handle overlay calling and to provide small areas of data storage for the communication of vital information between overlays. The resident area was kept as small as possible in order to allow the maximum amount of space for overlays.

A core map of the GCAD system is given in Fig. 4.1.

c) Overlay handling and storage

Since all system and user functions were to be called as overlays, the overlaying system had to be very fast. The system provided by DOS was not fast or versatile enough, so an independent high-speed overlaying system was written by Hamlyn [1]. This was later modified by Thompson [2] for greater efficiency and simpler operation.

An important requirement of the overlaying system was that overlays should be called just like subroutines; that is, each overlay does not have to specify which overlay is to be called next. This facility was provided by incorporating an overlay execution stack into the overlaying routines. This allows overlay calls to be stacked up ready for execution. A return from one overlay causes the next on the stack to be executed.

By stacking not only the overlay number but also an additional overlay segment variable, overlays can be segmented giving, in effect, multiple entry point overlays.

Overlays are stored on an overlay file on disc. This consists of a number of contiguous disc blocks with a directory occupying the first three. The overlay library program OVAL writes overlays, which may be of any length, into the best fitting contiguous space in the overlay file, and updates the overlay directory (see Fig. 4.2). It is also able to produce directory listings of the overlay file, eliminate the free space between overlays, delete overlays, and return free space available.

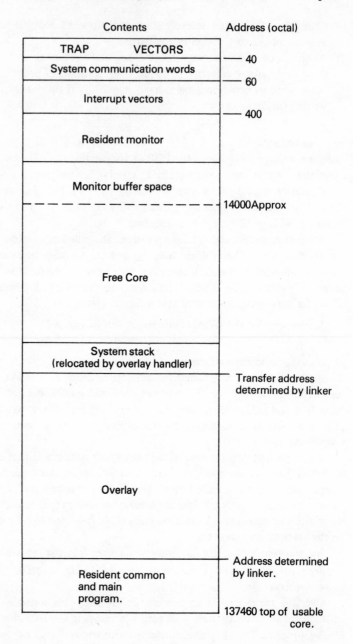

Fig. 4.1 – GCADS memory utilisation map.

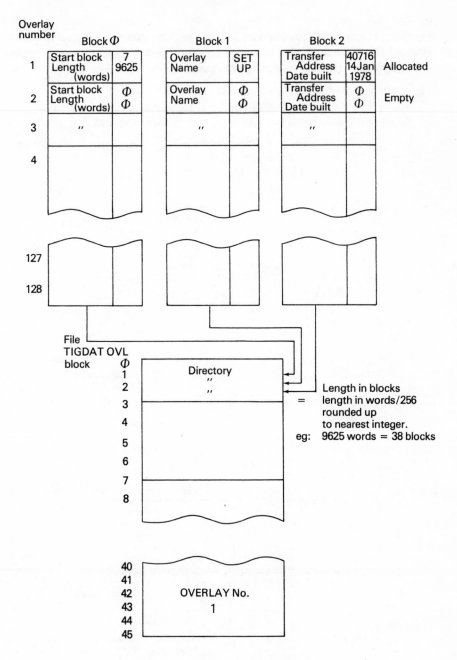

Fig. 4.2 – Overlay directory structure.

d) Disc data filing

For the operation of the basic system and user analysis packages it was considered essential that GCADS should contain a range of filing facilities varying in speed and complexity. Three types of file were specified (see Fig. 4.3):

1) Workfiles — high speed, random access, image format.
2) Semi-permanent files — medium speed, sequential access, binary format.
3) Permanent files — low speed, sequential access, ASCII format.

Each type of file can be written to or read from via Fortran calls to subroutines.

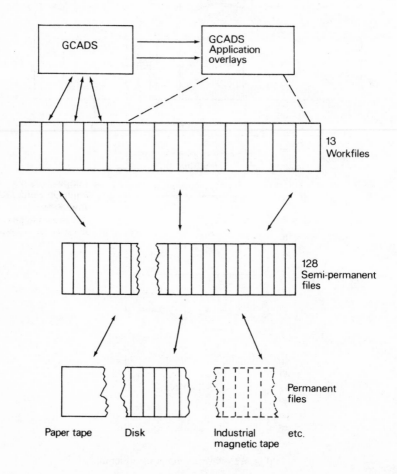

Fig. 4.3 — GCADS file system.

Workfiles are a set of 14 fixed length contiguously blocked files on disc. They can be written to at random without reference to the disc directory being necessary, and are therefore ideal for the temporary storage of volumes of data too large to be held in core.

Semi-permanent files are a set of 128 variable length random access files which are program deletable. It is intended that these should be used for the storage of workfile over periods of days or weeks.

Permanent files may be of any number on any medium: paper tape, magnetic tape, etc. They are intended as archival storage media for completed projects.

4.3.2 System modules

The software system consists of a number of system modules. Some of these have sockets into which user modules can be 'plugged' (see Section 4.5). The functions of the major modules will now be described.

a) Setup module

This module is called before any others when the system is run. It sets the following principal parameters:

1) Table origin   — The origin point of the table's coordinates (bottom left-hand corner).
2) Input origin   — The origin point of the drawing or drawing area,
3) Skew   to the tables x,y axis, defined by digitising two points representing a horizontal line.
4) Input scale   — The scale at which graphic information is to be input.
5) Grid factor   — Defines the lowest resolvable increment allowed
6) Text size   — The standard size for text added to a drawing.

Any of these parameters may be reset at a later stage by menu command.

b) User operations monitor (UOM)

When the setup routines are completed, the UOM overlay is called. This is the heart of the system and contains the interrupt, menu, level, and symbol handlers as well as the background loop'.

c) Background loop

The background loop has the following duties:

1) Read the table coordinates.
2) Convert them to real drawings coordinates: that is, subtract the drawing origin de-skew, scale and grid.
3) Apply CONTROL and TRAILING ORIGIN adjustments if in the appropriate mode, then convert to true 3-D coordinates.

4) Display the background messages.
5) Display 'pen' coordinates on the coordinate display.
6) Display a cursor on the screen related to the 'pen' position.
7) Look for a pen button interrupt.

Background messages are stored in an array in the resident common data area called MESSAGE. Five messages may be displayed, the first two being reserved for system use, and the rest for user programs. A user program can request a message to be displayed in background by writing the message to an appropriate part of the MESSAGE array.

d) Interrupt handler
The digitising pen has eight buttons (Fig. 3.2). When one of these is pressed it is detected by the background loop, and control passes to the interrupt handler in order to access certain facilities.

Buttons 2, 3, 6, and 7 perform the most commonly used functions which are break line, set 3-D trailing coordinate, control, and find. Button commands 2, 3, 6, and 7 are serviced by the UOM.

When button 1 is pressed the program determines whether or not the 'pen' is on the menu. If it is, then control is passed to the menu handler; if not, then a check is made to see if the system is in line mode. If it is, then the interrupt is serviced by the UOM; if not, then a return from the UOM is made to the next stacked overlay (which will be a symbol processor overlay (see 4.3.2g).

Button 4 is used to enlarge a portion of a drawing on the screen or plotting surface. It causes the display module to be stacked and the button 4 processing overlay to be called and executed.

When button 8 is pressed the user is invited to specify manually the next 3-D coordinate either as an absolute coordinate or as an offset from the last specified coordinate.

e) Menu handler
The GCADS menu consists of a 65 X 5 matrix of 15 mm squares located along the bottom edge of the digitiser, and a further 20 X 5 squares located along the left-hand edge. (see Fig. 4.4).

The overall menu is subdivided into a number of sections as follows:

a) System Commands consisting of:
    i) Display and manipulation of data (Fig. 4.5)
    ii) Data specification (Fig. 4.6)
    iii) Numeric input and calculator (Fig. 4.7)
    iv) Answer (Fig. 4.8)
    v) Joystick (Fig. 4.8)
    vi) Macros (Fig. 4.9)

b) User commands. (Fig. 4.10)

c) Files. (Fig. 4.9)

d) Symbols. (Fig. 4.11)

e) Macro library. (Fig. 4.12)

The user commands are similar in nature to the system commands except that they are written for user applications. For example, if the draughting system was used in conjunction with analysis programs then the analysis programs would be called from the user command area.

The commands are along the lower edge of the digitiser for ease of access. To permit flexibility in the arrangement of commands on the menu there is a mapping routine which translates the position of a square on the menu to a position on an abstract 'software' menu (see Fig. 4.13). The arrangement of the software menu is fixed, as its square number is an important parameter in the execution of many commands.

An identical technique is used to allow flexibility in the arrangement of symbol squares.

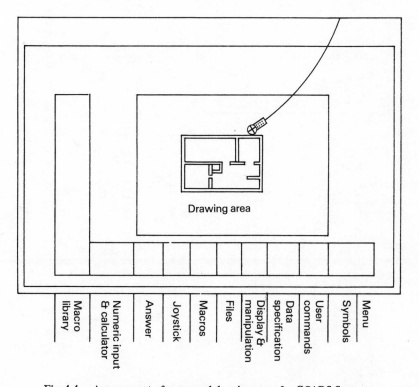

Fig. 4.4 – Arrangement of menu and drawing areas for GCADS System.

LINE EDITOR	PLOT PICT-ORIAL	CLEAR WKSP	DISPLAY PICT-ORIAL	X-Z DIRECT VIEW	SCALE UP OR DOWN WKSP	CURVE FIT DISPLAY	CONTIN MODE	ALPHA SIZE	LINE TYPE	PEN !	SET ACTIVE LEVELS
POINT EDITOR	PLOT ORTHOG	RESTORE WKSP	DISPLAY ORTHOG	Y-Z	SHIFT DATA AXIAL	CURVE FIT STORE	EXTRUDE SECTION TO SOLID	PERSPEC MODE	– – – –	2	SET INPUT LEVEL
SYMBOL EDITOR		TIDY WKSP		X-Z	ROTATE ABOUT SPEC AXIS	DISPLAY ALL MODES	REVERSE WKSP	I SOMET MODE	———	3	RETURN CURRENT LEVEL STATUS
DEBUG	REFIT PERSPEC WINDOW	ERASE SCREEN	SET PERSPEC WINDOW	ENABLE ORTHOG VIEWS	JOY STICK	DISPLAY ORTHOG ONLY	REVOLVE ABOUT SPECIF AXIS		———	4	DELETE LEVEL
	USE OLD WINDOW FIT	RESET WINDOW	RESET PERSPEC WINDOW	DISABLE ORTHOG VIEWS		DISPLAY PERSPEC ONLY			PRESENT LINE TYPE	PRESENT PEN	

Fig. 4.5 – A display and manipulation of data menu.

SET INPUT SCALE	SKEW CONTROL	SET PERSP PARAMS	SAVE RESDNT ON DISC	
SET OUTPUT SCALE	GRID FACTOR	SPECIFY TRAIL COORDS	RESTORE RESDNT	
2-D INPUT ORIGIN	Z DATUM MANUAL	Z DATUM COMPUTE		
DISPLAY ABS ORIGIN	DIREC COSINES	I SOMET ORIGIN		
DISPLAY TRAIL ORIGIN	FOCAL LENGTH	3-D ORIGIN		

Fig. 4.6 – A data specification menu.

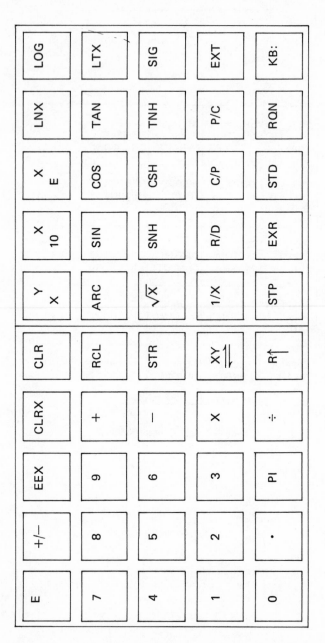

Fig. 4.7 – A numerical input and calculator menu.

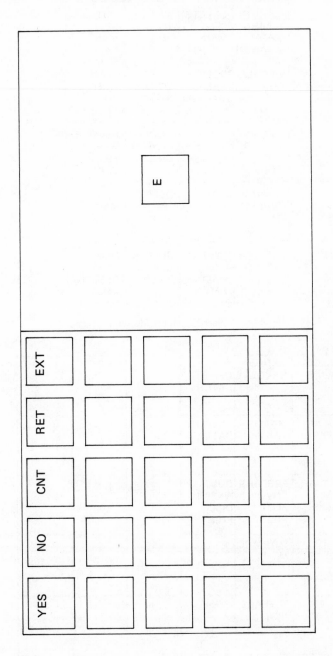

Fig. 4.8 – An answer and joystick menu.

CREATE MACRO FROM WKSP	DELETE SPEC MACRO	WKSP TO FILE	INIT NEW MAG TAPE
RECALL MACRO BY CODE	DELETE ALL MACROS	FILE TO WKSP	DUMP WKSP TO TAPE
DISPLAY MACRO DIREC-TORY		DISPLAY FILE OR DIRECT	RECOVER WKSP FROM TAPE
CHANGE DIRECT ENTRY		DELETE SPEC FILE	REWIND MAG TAPE
DISPLAY MACRO	SET UP MACRO TABLE	DELETE ALL FILES	SET TAPE RESTORE MODE

Fig. 4.9 — A macro manipulation and file menu.

CALCU-LATOR	AXIAL REVOLVE	SET UP BATCH			
POLAR WRAP	MATER TAKE OFF	EXECUTE BATCH			
PREPARE LIMIT FILES	PREPARE ENVOY PLOT FILE	DUMP WKSP TO DISC FILE			
EXTRA-POLAT	SURFACE MESH				
INTER-POLAT	PREPARE NEW MENU				

Fig. 4.10 — User commands.

LINE MODE	ARC MODE			DIMEN-SION MANUAL		TIGER SYSTEM 16/2/77
SPHERE MODE	CIRCLE MODE			DIMEN-SION AUTO		
RECT MODE	FILLET MODE			ANNO-TATE ───		
POLYGON MODE				ANNO-TATE ←		
ALPHA-NUMERIC MODE				ANNO-TATE ←──→		

Fig. 4.11 – A symbol menu.

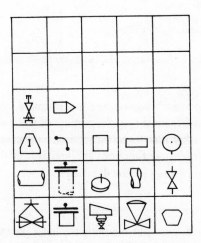

Fig. 4.12 – A macro library (only a small part is shown).

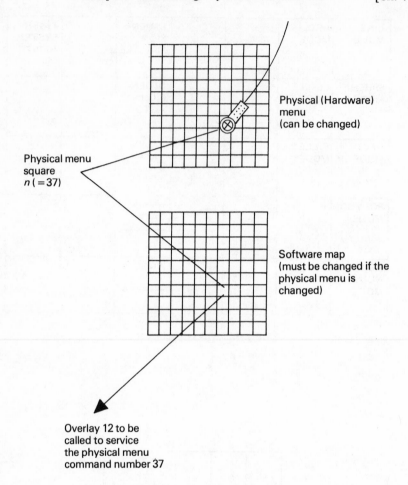

Physical (Hardware)
menu
(can be changed)

Physical menu
square
$n\ (=37)$

Software map
(must be changed if the
physical menu is
changed)

Overlay 12 to be
called to service
the physical menu
command number 37

Fig. 4.13 − Hardware to software menu mapping.

When a menu command is given, the menu square number is mapped onto the software menu, and from the square number of this menu, the menu handler determines which overlay is to be called to service the command. Commands are grouped by types, hence these functions are inrerlaced with ordinary commands.

When macro or symbol squares are digitised the appropriate handler is brought into operation.

f)   Level handler
Levels are used to divide up drawing data, so that a file can be broken up into a series of sub-files. Sixteen levels are provided in GCADS, and the level system is controlled direct from the menu.

A set of 16 bits form what will be referred to as a level 'sieve'. All operations on drawing data are carried out after it has passed through the sieve. Each bit in the sieve represents a level. If the bit is zero, the operation is not carried out on that level of data and vice-versa. Therefore individual levels can be selected for displaying by setting the appropriate bits in the sieve.

When a display level is specified the sieve bit corresponding to that level is set. The command SET INPUT LEVEL enables the user to select a new data level. Menu commands exist to open or close all holes in the sieve.

g) Symbol handler

All graphic items which are not straight lines are handled as symbols. Each symbol is processed by a number of symbol overlay segments called:

Symbol initialiser segment.
Symbol point processor segment.
Symbol display segment.

When a symbol square is digitised the symbol initialiser overlay segment is stacked. When, subsequently, a point is digitised by using button 1, the initialiser is entered. This has the job of storing the digitised point, stacking the symbol point processors, and returning to the UOM. Subsequent points digitised will cause the symbol's point procrssors to be executed until the symbol is complete, when the symbol display segment will be entered to display the symbol on the screen.

Symbol overlays can be written by applications programmers, and can serve as construction techniques as well as a means of entering non-standard graphic items, such as ellipses or rectangles.

h) File handler

Workfile 1, generally referred to as the WORKSPACE, holds data for the current drawing before it is transferred to one of the 128 semi-permanent files available in the system. These 128 files are accessed by number, and the user has the ability to write a file, recall an old file to the workspace, display the contents of a file on the VDU, or delete a file. If the user tries to write over a file that is already assigned, the previous contents are lost, so for security of important data it is possible to 'protect' a file so that it cannot be overwritten until the protect state is removed. In addition, a 'user' directory is available in which an appropriate description of the contents of the file can be stored. This directory is merely for convenience and is not essential to the operation of the basic file handler.

i) Macro handler

A further set of 128 files is available for storing macros (defined in Section 5.6). For speed of operation these are accessed by a block of squares on the left-hand edge of the menu. Unlike files, the 3-D macros have to be specially prepared before being filed in order that they may be recalled with the maximum efficiency. During drawing input, a macro may be recalled simply by digitising over the appropriate square. In addition the macro directory contains a codeword which may also be used for recalling the required macro. The macro handler displays the macro in perspective in the top right-hand corner of the screen and marks the 'reference points' the user must digitise to place it in 3-D space. From these points the handler works out the required scale factors, rotations, and translations and places the macro accordingly. In some examples, such as, valves, the macro may be defined by a set of reference tables depending on, for example, the pipeline diameter. In this case it is possible to store such a table; the macro handler will extract the relevant data for the selected macro and manipulate it accordingly, using the digitised reference points merely for orientation.

j) Display and plot modules

The function of these modules is to display or plot the data in the workspace. Symbols are stored with various identification markers, and each module contains a section to display any of the symbols available. This is considerably faster than the existing system of stacking the relevant symbol overlay to the display section.

4.4 DATA STORAGE

One of the first decisions made in designing GCADS concerned the type of structure and format to be used for the storage of drawing data. Several types of data storage structures were examined, and some test programs were written.

In early systems data was stored as a list of codes in the order in which they were entered. This meant that many points having two or more lines connected to them appeared in the data more than once. It seemed likely, therefore, that storage space could be saved by storing points separately from their interconnections, with one file to store point coordinates and another to store interconnection data (see Fig. 4.14).

A test program for this data system was written, with lines as the only graphic items. The system worked well if the point file was small enough to be largely core resident. However, when a significant part of the point file was stored on disc, the system became slow because of the number of disc accesses needed to retrieve points. The reason for this can be seen from the following example.

Suppose that a drawing consists of 700 lines connecting 300 points. Then as a line connects two points there must be at least 100 duplications of points. Taking the display time for one vector as being 3×10^{-3} seconds (3 ms), the drawing would take 2.1 s for the vector-generating hardware to display.

If a point takes two variables to define its x,y coordinates, and an interconnection also needs two variables, there are 600 variables of point data and 1400 of interconnection data.

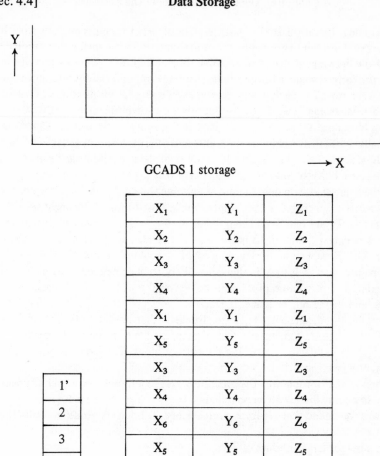

GCADS 1 storage

Separate Point Storage

Interconnection Store

(' Indicates the start of a line)

Point Store

Fig. 4.14 – Example of separate point storage.

Assuming, therefore, a 600 variable area of core for points and the same for interconnections, four disc reads of approximately 35 ms each are necessary to display the drawing, giving a total display time of 2.24 s.

In the early systems without separate storage of points and interconnections, points were stored in such a way that each line required four storage variables, giving a total storage volume of 2800. With the same buffer size of 600 variables this would require 5 disc reads to display, giving a display time of 2.275 s. This is marginally worse than when storing points and interconnections separately. However, by storing them separately the amount of data required has been reduced by 28% from 2800 to 2000 variables.

If the amount of data held in core at any one time is now halved it is possible, in the worst case, that the 100 points that are used more than once are not in core when they are required. The number of disc reads necessary to display then becomes between 7 and 107 for the separate storage system and 10 for the list system. This gives a total display time of 2.275 to 4.65 s and 2.35 s respectively. The separate storage system is therefore likely to become much slower than the list system as the volume of data grows, especially as on a small machine core storage, and hence buffer space, is scarce.

Another disadvantage of separate storage is that double buffering of data is virtually impossible. This means that it is impossible to save time by reading the next block of data while processing the current block, as the data is not accessed sequentially (see Fig. 4.15).

When searching for points the separate storage method is obviously the faster, as only the point file has to be searched.

It was concluded that separate point storage, although it produced reductions in data volume, was

a) slow for large volumes of data,
b) complicated and difficult for applications programs to handle,
c) inflexible, as different data types were difficult to add.

It also depended upon a number of points being used several times. This often will not be the case.

It was therefore decided that data should be stored in a list type structure.

Another method of economising on data volume is to store macros (repeated sequences of data) once only. This means that macros would be stored on a 'drawing macro' file and referenced by a code in the data list specifying not only the macro used but its scale rotation, position etc.

This system is similar to the separate point storage system. It does not, however, suffer from the same time problems as separate point storage, as macros are usually quite large, and in any case they require a disc read each time they are accessed.

The disadvantages of this system are:

1) Editing the contents of a macro when it has been positioned on a drawing becomes difficult.

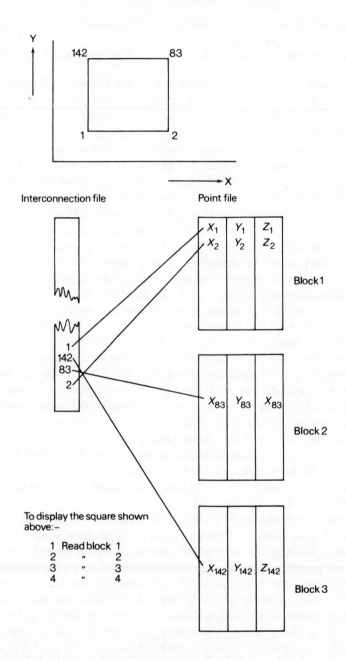

Fig. 4.15 — The possible number of non-sequential disc accesses produced by a separate point storage system.

2) Display time is slightly increased as a transformation has to be applied to the macro before it is displayed.
3) Finding a point within a macro requires a complicated searching routine.

For these reasons it was decided that in the general CAD system macros should not be stored separately, although it is recognised that applications where many standard items are used could use such a system to great advantage (for example, electronic circuit design).

Symbols were defined as being program-generated sub-pictures; that is, the symbol data is used to call a section of program to generate the apecified symbol picture. It is therefore possible to treat symbols in the same way as lines and to store them in the data list. It was therefore decided that data should be stored as a simple list of 3-D records. These records consist essentially of a 3-D coordinate, X, Y, Z, and an I code whichspecifies the meaning of the coordinate; for example, $I=1$ means that this is the start of a line. Since PDP 11 Fortran assigns two words of storage to both real and integer variables (although in the latter case one word is not used) it would be possible to store 32 such records per block (256 words), the smallest addressable area of store on the disc. Old 2-D systems employed such an approach: 40 X, Y, I records were stored and 16 words 'wasted'. Although this was inefficient, it did mean that data was easily accessible with Fortran programs. However, experience with this method has shown that data can become somewhat cumbersome and difficult to manipulate, so a radically new approach was tried. A set of subroutines were written to handle the reading and writing of data from and to the workfiles. Since the user's only communication with the data file is via these routines it was possible to eliminate the 'wasted' integer word and so store 36 3-D records per block. Records are input and output in 'auto-incremental mode'; that is, the system automatically moves on to the next record after dealing with the current data. At any time the current position in the file is defined by the input and output pointers, which are completely independent of one another and may be set by the user to any value to facilitate truly random access of the data.

In earlier systems special records were inserted to indicate, for example, a change of input level. This system worked well in practice only if data was accessed sequentially. If a user wished to access a record in the middle of the data file it was necessary to search all preceding records to find the relevant level, line type etc., and if this were true for the 3-D system the advantage of the greater flexibility obtained so far would be lost. This problem was tackled in earlier systems by maintenance of a level index, indicating the start and end position of each level in the data file. This system was very difficult to maintain. For example, if a point was deleted in the database the whole index had to be re-written — a time-consuming process. The most flexible approach is for each point to have its associated level, line type etc. included in the database with the I, X, Y, Z records. If standard PDP 11 Fortran is used, this approach necessitates the use of many more words of storage and so increases greatly the size of the overall database. However, with the introduction of the previously mentioned routines to handle the data it was found possible to pack the required data into only one word and to combine it with the I code associated with each point. By limiting the range of some parameters it is possible to store them in only a few bits of a word and so store several of them in the same word. The following ranges were selected:

I CODE	$0 \rightarrow 31$	=	5 bits
LEVEL	$0 \rightarrow 15$	=	4 bits
PEN NUMBER	$0 \rightarrow 3$	=	2 bits
LINE TYPE	$0 \rightarrow 3$	=	2 bits
DISPLAY CONTROL	$0 \rightarrow 3$	=	2 bits

If the level is stored as, for example, $0 \rightarrow 15$, the level returned to the user is in the range $1 \rightarrow 16$ to facilitate easier manipulation with Fortran programs. Pen number and line type are treated in the same way. The current input parameters are stored in a Fortran common block/CODES/; the record handler takes its data from this area and encodes it whenever new data is entered. When the user is reading from the data file the handler decodes the above data and places it in another common block/CODOUT/. In this way it is easy for the user to access these parameters by using standard Fortran programs.

Although the system has 13 random access files for data storage, it is not practical to have all 13 accessible in the above manner at any one time, since each input and output file requires a 256-word core buffer. It was found that 3 sets of buffers are adequate for most uses. Of these, the first is generally assigned to the workspace; the remaining two are 'scratch' buffers used by programs needing temporary data storage separate from the workspace. It is possible, by subroutine call, to allocate any of the 13 files to a particular buffer, thus the only restriction on the user is that not more than 3 files can be used for input and 3 similar files for output at any given time.

Symbols are generally stored in the workspace with the minimum amount of data necessary for the system to define them completely. For example, a sphere is stored as a centre point and a point on the surface only; from these the display and plot routines generate the appropriate projections. All symbols follow these conventions:

The first and last records of the symbol are indicated by an I code of 3 and 4 respectively, the X, Y, Z coordinates indicating the high and low limits on the symbol respectively. This information is useful to the symbol editor and to the display routines to decide whether the symbol is within the current area of interest.

The second and third records are indicated by an I code of 5 and carry data to identify the symbol type and its length, and must therefore obey a standard format. The second record is for the user to include whatever extra data may be necessary for that symbol, and is not used by the system software. 3-D macros are stored in the same manner except that they are bounded by I codes of 13 and 14, since it is possible to have a symbol within a macro, and some confusion may result from the 'nested' use of 3 and 4.

4.5 COMMUNICATION WITH APPLICATION ORIENTED SYSTEMS

In general, any CAD system designed for a specific application will require graphics facilities peculiar to the application. Applications programs will also need to interact directly with the user and with GCADS graphic data files. These facilities are provided in GCADS by:

a) User menu commands.
b) User symbols.
c) The system subroutine library.
d) Graphic data file-handling routines.
e) System COMMON storage.

To provide extra graphic facilities the application's programmer can write symbol overlays to the specification of the system. These overlays are 'plugged into' the GCADS system by editing their menu position into the software menu map, including the relevant display and plot routines in subroutine DISALL and overlay PLOTOV respectively, then rebuilding the background overlay (DIGOV), display overlay (DISALL), and plotting overlay (PLOTOV).

Application overlay programs can be written in standard Fortran IV with the exception that CALL OVRETN must appear instead of the Fortran STOP statement. An application program is brought into operation from a menu command or by being called from another application program. To 'plug' an applications program into a menu command the subroutines MENMAP must be edited and the monitor overlay (DIGOV) rebuilt.

The system subroutine library contains a collection of subroutines which may be used by applications programs running under GCADS to communicate with the user, handle data, etc.

Any data inserted with an I code other than those used by the system will be ignored by all system routines, thus the user is able to insert special codes applicable to his programs alone which will not affect the general running of the system. All data with an I code of 31 is assumed to be null data; it generally results when a line or symbol has been deleted from the workspace. This data is of no further use and may be removed by overlay TIDY.

GCADS workfiles numbers 2 to 11 are available for use by application programs.

Various core resident COMMON variables are available for reference by application programs. In addition a 256 word COMMON/SCRACH/ area is available solely for use by application overlays, providing an easy means of inter-overlay communication.

REFERENCES

[1] Hamlyn, A. D. *The application of CAD techniques to building engineering design.* PhD thesus. University of London 1974.

[2] Thompson, D. *Computer Aided Draughting in Engineering.* PhD   thesis. University of London 1978.

CHAPTER 5

Operation of the GCADS Draughting System

A description of the facilities available within the draughting system, with procedures for using them, is now given. These facilities can be controlled from a menu, a keyboard, or a combination of both. The operator can decide how the various functions are to be controlled. The menu system enables the user to operate the system with very little training, since each instruction is clearly written within each menu square. However, using a menu for every instruction does mean that the operator must constantly move to and from the position where he is working with the menu. Using the cursor, with the keyboard shown in Fig. 3.2, overcomes some of these problems.

The system now described can be operated with a cursor having a single button or one with a keyboard of eight buttons.

5.1 SETTING UP THE SYSTEM

To run GCADS the following procedure should be observed:
a) Bootstrap the GCADS system disc in system drive.
b) Load a data disc in disc unit DKI.
c) Type LO 2,2 followed by a carraige return.
d) Type RU GCADS followed by a carriage return.

The message ZERO TABLE will then be displayed on the screen. The table is zeroed by the operator digitising the extreme bottom left-hand corner of the menu. This sets the coordinate reference point for the digitising table.

The operator is now free to use any of the GCADS commands described below.

All initialising conditions are set by default in the set-up procedure. These can be changed at any time by means of the following menu commands:

a) SET OUTPUT SCALE
   The output scale determines the size to which the drawing is to be plotted. Output scale is set equal to the input scale in the initial set-up. The output scale is set by the operator in putting its value from the keyboard area of the menu.

b) 90° LOCK
    15° LOCK
    Two CONTROL modes are available in GCADS. In CONTROL mode
    lines are constrained to an angular grid. The grid can be 15° or 90°. A
    90° grid is set by default. This constrains lines to be parallel to either the
    x or y axis.

c) TRAILING ORIGIN
    ABSOLUTE ORIGIN
    2-D table coordinates displayed on the screen and coordinate display
    unit may be relative to the last point digitised (trailing origin), or relative
    to the drawing input origin (absolute origin). Trailing origin mode is set
    in the initial set-up procedure.

d) LINE TYPE
    The line type is set to solid initially.

e) PEN
    The pen number defines the pen to be used for plotting. By using four
    pens different line thicknesses can be drawn. The default pen number is
    pen 1.

f) SET INPUT LEVEL
    The level at which data is being added to a drawing is defined by giving
    the SET INPUT LEVEL command and digitising a level number. All
    subsequent data added to the drawing will be assigned to this level until a
    new input level is defined. The default input level is number 1 (the
    COMMON level).(See Section 5.4).

5.2 GRAPHIC INPUT

Digitising may take place in full three dimensions. Initially the drawing area is
divided into four main areas correspondingly shown on the display screen. These
four areas correspond to four orthogonal planes so that a design can be developed in more than one orthogonal view. This is essential for the representation
of three-dimensional objects. Fig. 5.1 shows a solid object and four orthogonal
views of the object. The four areas of the workspace are A, B, C, and D, and these
represent planes x-y, z-x, z-y, and y-z respectively. If a point is digitised in one
view it must be accompanied by a second point digitised in one of the three
remaining views in order to specify the point in three dimensions.

When the system is operating in 3-D mode the plane in which the coordinates
of the digitising cursor are determined are automatically converted into three-
dimensional coordinates. Thus the z coordinate of an area in the x-y plane may
be digitised in the z-y or z-x planes followed by all the coordinates necessary to
specify the area in the x-y plane. This avoids much repetitive digitising of z
coordinates, thus it is only necessary to digitise in a second plane when the third
dimension changes.

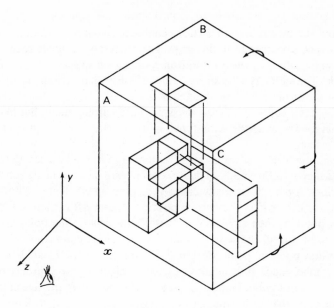

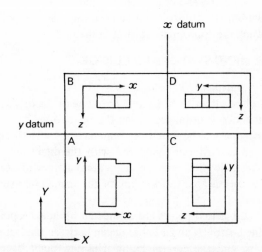

Fig. 5.1 – Various projections from a solid object.

To separate the planes a point known as the x-y datum (ZX, ZY) must be defined before any data input. This point corresponds to the common vertex of the planes. If it is defaulted at the initialising stage, the table is then divided in four equal areas. A permanent cross on the screen, indicating the datum point, is moved to the top right-hand corner of the workspace, and the system is then in 2-D mode with the coordinates digitised in the x-y plane only.

The system can be operated with a cursor or pen with a single button switch and the menu, or with a cursor containing eight buttons in the form of a keyboard and the menu. If the single button system is used, then functions which can be called on the eight button system are displayed on the screen with a pointer. The point can be moved down the list of functions by pressing the digitising button until the desired function is selected. The operation of the system now to be described will refer to the eight button arrangement, but in each case the reader can imagine the same function displayed for selection by the single button system.

Graphic input takes place in one of a number of symbol modes, the basic one being line mode In this mode all digitised points are taken as being connected by straight lines. A point is normally digitised by pressing the digitising button when the digitising cursor is in the correct position. A line AB is therefore drawn by digitising point A, then digitising point B. Digitising a further point C will cause a line BC to be drawn, and so on. If button 2 is pressed before a point is digitised, no line is drawn to that point. Button 2 is therefore used to 'break' a line.

The symbol mode to be used is selected by digitising the appropriate symbol mode square. In all symbol modes messages are displayed on the screen informing the operator of the significance of the point he is about to digitise (whether the point is the centre of an arc, end of a dimension line etc.).

To aid in the input and construction of a drawing, various facilities are made available by digitising the appropriate squares in the menu.

5.3 USE OF THE MENU SYSTEM AND CURSOR

5.3.1 Data input — buttons 1, 2
Data input takes place either in line mode or in one of the other system modes. The current input mode is displayed in the top right corner of the screen where most system messages are to be found. If no mode is specified then the system is in line mode. When in an input mode and with the digitising cursor over the drawing area of the table (not over the menu area) a tracking cross is displayed on the screen. The user is able to issue any of the menu commands while in an input mode.

In line mode all digitised points are connected by lines. A point is normally digitised by pressing button 1, and a line is drawn between the last point digitised (the trailing origin) and the current point. The new point becomes the new trailing origin, and the cartesian coordinates will be set to zero at the new point.

To break the series of lines produced by digitising several points with button 1, button 2 should be pressed. After button 2 no line will be drawn between the current trailing origin and the next point digitised. Thus button 2 is used to 'break' a line. If button 2 was the last button pressed, then a message LINE BROKEN will be displayed in the top right corner of the screen. Alternatively, a break in line mode can be achieved by calling BREAK LINE on the menu ANSWER area, using button 1.

A symbol mode is entered by digitising the appropriate symbol mode menu square. A message will indicate which mode is in use, and a further message will inform the operator of the significance of the next point to be digitised (for example, the centre of a circle or the start point of a text string).

To aid the data input and construction, several utilities are available and can be called by pressing an appropriate button or by selecting the command on the menu 'ANSWER' area.

Set trailing point — button 3
As previously described, a fully 3-D coordinate is entered by digitising in two orthogonal views. Button 3 allows the trailing coordinates, for example Z when digitising in x,y plane, to be reset without entering a point in the workspace. The message SET TRAILING POINT is displayed, and the user must then digitise the appropriate coordinate. This facility is useful when finding a point in 3 dimensions and placing macros.

5.3.2 Windowing — button 4
When the User Operations Monitor is entered the screen represents the whole drawing area of the digitising table. Since the screen is only one twelfth the size of the table, the data is displayed at a much smaller scale than that at which it is input. Thus details are not clearly visible. To overcome this problem the user can select an area of the table which he wishes to view at a larger scale. The procedure is:
a) Press button 4, or call WINDOW utility from the menu. A message LEFT HAND WINDOW POINT is displayed.
b) Press button 1 to digitise the bottom left-hand corner of an imaginary rectangle that will cover the desired area of the table. A message RIGHT HAND WINDOW POINT is displayed.
c) Move the digitising cursor from left to right across the table towards the bottom right cormer of the imaginary rectangle. As this is done a rectangle is displayed on the screen. When this is seen to enclose the required area, digitise the right-hand window point. The data within the rectangle will then be displayed on the full area of the screen.

The menu command RESET WINDOW is used to return to the initial situation in which the screen represents the whole drawing area of the digitising table.

5.3.3 Angular control — button 6
In control mode, lines are constrained to be at angles which are a multiple of $90°$ or $15°$ to the x-axis depending on whether $90°$ LOCK or $15°$ LOCK was the last LOCK menu command selected. When control mode is set, a message is displayed with the other system messages in the top right cormer of the screen either CONTROL 90 MODE or CONTROL 15 MODE. Control is switched on and off by successive presses of button 6. CONTROL MODE can also be called from the menu by button 1.

5.3.4 Finding – button 7

The FIND facilities of button 7 make it possible to join new data with existing data and to position the trailing origin accurately at a known position.

Pressing button 7 causes the message DIGITISE NEAR POINT to be displayed on the screen.

If the user wishes to join a line to an existing point or to find a point for placing a macro, etc. then the cursor is moved close to the point and button 1 is pressed. If there is no point within a 5 mm circle centred on the pencil position, the message NO NEAR POINT is displayed on the screen for a short time.

In summary, the main uses of the FIND facility are:

a) To ensure that data joins up correctly.
b) By using a break line before finding, to position the trailing origin at a known point in order to:
   i) Use it as a starting point when entering drive mode.
   ii) Measure distances.
   iii) Use the position as an origin for a macro.

5.3.5 Drive mode – button 8

It is often necessary to specify exactly the location of a point. This function allows the user to define the position manually from the numeric input area of the menu. The point may be either relative (that is, as an offset from the last defined position) or absolute (that is, relative to the drawing origin), the mode of operation being selected after pressing button 8. This feature will also be found useful in conjunction with button 3 to specify an exact trailing coordinate when inputing a drawing from the digitiser.

5.3.6 Floating assignment – button 5

Since it takes much longer to move the pencil to a menu command square than to press a button, it was decided to give the user the facility of choosing which menu command he would most like to be able to call merely by pressing button 5. Button 5 is set up so that it is equivalent to the command LINE MODE. Thus when the user wishes to return from a symbol mode to line mode he need only press button 5.

To assign button 5 to another menu command, position the pencil over the required menu command and press button 5. Button 5 is then equivalent to that menu command.

5.4 LEVELS

The last facility, an important factor for the operator, is the division of data into different levels. The operator can select one of sixteen different levels under which to input data, by digitising the SET INPUT LEVEL menu command followed by a level number. The operator is able to specify levels to be active or dormant. Data under an active level will be displayed and processed by the system. In general, only filing operations will be carried out on data under dormant levels. Thus by storing data under different levels and by declaring them active or dormant the operator is able to:

a) Reduce display and processing time.
b) Reduce editing time.
c) Obtain clear views of different categories of data: for example, to distinguish between water pipes and electrical cables.

To make the best use of this facility the user should sit down and think about the best way to split his particular data into levels. Much time can be saved by classifying data under different levels before starting a job.

5.5 FILING AND DISPLAY

All data entered in one of the input modes described in Section [3.2] is automatically buffered into a random access file called the 'workspace'.

On entry to the system the workspace is empty. Data left in the workspace at the time a run is terminated will be lost. The operator is able to clear the workspace by giving the CLEAR WORKSPACE command. The menu command ERASE SCREEN clears the screen of any information which has previously been stored on it, and the workspace is displayed by digitising the DISPLAY WORKSPACE command.

The data in the workspace may be stored under different levels. To display all levels of data the user must digitise the menu command DISPLAY ALL LEVELS. To display only selected levels of data the user must digitise the command SET DISPLAY LEVELS followed by the level square numbers of the levels that the user wishes to be displayed.

The data contained in the workspace can be filed in any of 128 semi-permanent mass storage files, which are accessed by number. There are five menu commands associated with these files. After digitising one of these menu commands the user additionally digitises a file number to indicate which file is to be operated on. The five commands are:

1) WORKSPACE TO FILE
The file selected, if not a protected file, is overwritten by the contents of the workspace. The user is instructed to make an entry in the user directory: if he does not, a default entry is made.

2) FILE TO WORKSPACE
The contents of the selected file are added to the workspace in the current input level and also displayed on the screen. A message FILE IS EMPTY is issued if the selected file is not occupied.

3) DISPLAY FILE OR DIRECTORY
The content of a specified file is displayed on the screen without being added to the workspace. Alternatively, a listing of the user directory is produced on the screen.

4) DELETE FILE
The specified file is deleted. But if it is protected, an operator confirmatory response is requested before the operation is completed.

5) DELETE ALL FILES
All files are deleted and the directory cleared after a confirmatory response from the operator, irrespective of the protection specified for individual files.

The above procedure is useful for storing small files that may have to be accessed quickly. Since the physical size of a disc limits the size of files that may be stored, it is obviously necessary to have some form of bulk medium for the storage of large data items or the archiving of complete drawings. A cheap and effective method is to use magnetic tape. One tape can hold many drawings and can be stored compactly when not in use. The chief disadvantage with this method of stotage is the time taken to retrieve a drawing, particularly towards the end of a tape.

The following menu commands control all tape storage operations:

1) INITIALISE MAGNETIC TAPE
This command is obligatory when a new tape is loaded before writing data. It records the number by which this tape is to be known and effectively deletes any previous data by writing end of file narkers at the start of the tape.

2) DUMP TO TAPE
The content of the workspace is written as the next file on the tape, which is then left ready to accept further data should this command be used again.

3) RECOVER FROM TAPE
A file is read from the tape and written to the workspace. Each file in a particular tape is known by number, and the system can be made to retrieve a numbered file out of sequence or to step through the tape retrieving each file sequentially.

4) SET RESTORE MODE
In DEFAULT RESTORE mode, the tape file is added to the existing contents of the workspace. The procedure is similar to that used for the semi-permanent disc file handler. The user can select OVERWRITE mode, where the old contents of the workspace are lost when a new file is read. This mode is somewhat faster and is useful when paging through files to locate a particular drawing.

5) REWIND TAPE
The tape is rewound to the first file. This facility can be used to advantage if one of the first files is required and the tape is positioned some distance past it. The system will, of course, rewind automatically when told to RETRIEVE THE FILE; but by specifying the rewind command earlier, the

operator can be doing something else while the tape is rewinding. Whereas when the tape is rewinding to search for a file the machine is dedicated to that task alone.

It is also convenient to be able to dump the contents of the workspace to any device or file so that, for example, calculations of stress may be carried out on another machine where the data may have to be transferred by paper tape. A simple routine exists to write the data, as I,X,Y,Z records, to any specified device either as ASCII strings or in unformatted binary mode. The former is indeed obligatory when an output device such as the line printer is selected, and where the operator may be required to check the data manually. The latter has the advantage of being more compact, but can only be read by a computer.

5.6 MACROS

Macros are standard components or items that are in regular use. A menu area, of 100 squares, called MACRO LIBRARY has been set aside for filing macros and recalling them. Although only 100 squares appear on the menu for the library, the macro library squares may be paged to give a large number of squares. Alternatively, macros may be filed under a code and recalled by the same code.

The macro facility greatly enhances a CAD system in that similar items, components, or sections of data may be created in the workspace only once, filed, and subsequently reproduced with little effort.

The user must study a problem in advance and decide if groups of data can be treated as macros, since it is repetition which can be exploited in CAD.

A set of menu commands controls the macro handler. When a new macro is to be included the user must first digitise it in the normal manner and then select the CONSTRUCT MACRO command. All macros must have a standard format, and this routine will add the necessary data. In general, the placing of a 3-D macro on recall is controlled by 3 digitised points, the minimum number required to completely define the position in 3-D space. They must correspond to 3 predefined points in the macro, and the first task when setting up a new macro is to define these positions. The system is slightly less flexible than allowing the user to define the reference points when the macro is recalled, but is very much more efficient and speedy when it comes to digitising a large drawing. Having defined the reference points, further information may be requested, depending on the application for which the macro is intended, for example pipe routing. The operator may be requested to specify if a value is reversible — if it is not, and the operator tries to place it backwards at a later stage, an error message will result. Since macros are recallable by code the operator is requested to specify a codeword and a brief title, and finally to digitise over the square corresponding to the required macro file number. All macros are automatically protected files and cannot be inadvertently overwritten.

Many standard components vary only in size between applications, and these sizes are predefined in a reference table. For example, a value may be of fixed design, but of varying size depending on, say, the pipeline into which it is to be fitted. A facility exists for storing the appropriate reference tables, and when the macro is subsequently called the system can look in this table and extract relevant data. A menu command SET UP TABLE enables the user to input such a table by typing the relevant entries at the keyboard.

Macros can be recalled either by digitising the square RECALL BY CODE, then typing the codeword on the keyboard, or by simply digitising over the appropriate library square. The macro is then displayed in perspective view in the top right-hand corner of the screen, and the user is invited to place the reference points, which are marked by small crosses for clarity. If the particular macro is not defined by a reference table, these points will be used to obtain scale factors and orientations, otherwise the appropriate sizes will be extracted from the table and the digitised points used merely for orientation.

When the user had supplied all the data required by the macro processor, the appropriate scale factors and rotations are applied. Finally, the complete macro is displayed in the correct position on the screen and is added to the workspace. The system recognises macros as symbols much as circles and arcs, therefore they can be removed with one operation if required. Additional information regarding macros is stored for use later to produce a material take-off from the complete drawing.

As with the normal files, the user has the option of deleting a macro file, deleting all macros, or displaying the directory on the screen. The user also has the option of changing the directory entry without having to reconstruct the macro.

5.7 EDITORS

There are four editors: point editor, line editor, macro/symbol editor, and level editor.

5.7.1 Point editor

The point editor is used to move points in 3-D space or to delete redundant points. The routine is accessed by the POINT EDITOR menu command.

The routine functions are in either orthogonal or perspective mode. The user is asked to place the cursor near the relevant point as it appears on the screen. In orthogonal mode the point can be found from its projection in any of the 3 views available. The system then searches for the relevant point, and it is marked by a small cross on the screen for verification. The point may be accept-ed by pressing any button on the cursor, or disregarded by moving away from it, in which case the system will start searching for another point. Having accepted a point, the user may either delete it entirely or reposition it by digitising its new position. In either case the relevant modification is made. The system then starts searching for a new point until told to exit. The screen is then erased and the modified workspace displayed, the system returning to background mode.

5.7.2 Line editor

The line editor functions in a similar way to the point editor except that the proximity of the cursor to a line rather than a point is checked. When a line is located it is flashed on the screen with the message ANY BUTTON TO DELETE. Pressing any button removes this line from the database. If the user does not wish to delete this particular line, the system will automatically start searching for a new line if the cursor is moved away from the current line. The EXIT command reverts to background mode after erasing the screen and displaying the modified workspace.

5.7.3 Symbol and macro editor

Symbols and macros may contain many lines and points, and it is important to be able to delete these as one unit. Since macros are defined in a similar fashion to symbols it is possible to use the same routine to remove either type. Two modes of operation of the symbol editor are possible:

1) *Random search*

   This routine makes use of the first and last records of the symbol which contain its low and high limits respectively. As with the other editors, the user must place the cursor so that it lies within the projection of the required symbol in whichever view he chooses. The limits of each symbol are checked until the appropriate one is located. Its boundary, rather than the symbol itself, is flashed on the screen with the message ANY BUTTON TO DELETE. The search process is continuous, and if the cursor is moved the system will start searching for a new symbol. A deleted symbol is flagged in the work-space as null data but to save time it is not removed at this stage. It should be removed by overlay TIDY at a later stage.

2) *Sequential search*

   In this mode the system offers up every symbol in the workspace for deletion sequentially; the user specifies YES or NO from the ANSWER area of the menu, depending on whether the symbol is to be deleted or not. This process is cyclic; if the routine reaches the end of the data it returns to the beginning until the user specifies EXIT, when the system returns to background mode after redisplaying the modified workspace.

5.7.4 Level editor

The user may choose to delete all data in a particular level. This is useful if the user has planned his drawing correctly and has digitised sections in different levels. If one section turns out to be completely wrong, this is a much faster method of deleting it than using the symbol and line editors to remove each data item individually.

5 8 MISCELLANEOUS FACILITIES

5.8.1 Continuous mode
The continuous mode facility enables the user to digitise smooth curves. In this mode the position of the cursor is tracked, and whenever it is moved more than a certain distance from the previous point a new point is entered. For obvious reasons, continuous lines can only be drawn parallel to one of the major planes, but is it possible to reset the trailing coordinate within the routine by pressing button 3. Any other button will start or end point input, depending on the state at the time of pressing. Continuous lines are entered in the workspace as symbols for ease of removal if this is necessary.

5.8.2 Plotting
Any drawing can be plotted in the on-line flat-bed plotter by digitising the PLOT PICTORIAL (perspective or isometric) or the PLOT ORTHOGONAL square. The user must then specify the size of paper to be used and the required plotting scale. He is then given the option of allowing the system to place the drawing centrally on the paper or to place it manually from the digitiser. In the latter case a rectangle is displayed on the screen indicating the area of the paper, The rectangle is moved about with the cursor, and finally placed by pressing a button. Any data which falls outside the area of the paper is automatically 'scissored' so that the drawing does not run off the paper.

5.8.3 Debug
The menu command DEBUG enables the applications programmer to examine any of the semi-permanent files. The casual user need not be concerned with this command.

5.8.4 Curve fit
It is possible to fit a cubic spline curve to all of the linked points in the work-space by the use of this menu command.

The following messages are displayed:

CURVE FIT WORKSPACE
B1 DISPLAY ONLY
B2 STORE
B8 EXIT

Pressing button 1 will display the curve. The workspace is unaltered.
Pressing button 2 will store the curve in the workspace as many short lines.
Button 8 allows an exit from the curve fitting command.

5.9 THREE-DIMENSIONAL DISPLAY FACILITIES

When working in three dimensions the Z datum normally divides the workspace into four equal areas. The datum may be moved by digitising Z DATUM followed by digitising the new position, or computed by specifying Z DATUM COMPUTE, then digitising the projection of two points in two of the orthogonal views. For working in two dimensions, the Z datum is simply moved off the workspace at the top right-hand corner.

When three-dimensional data is built up in the system, there are a number of facilities which greatly assist in viewing the data for examination, editing, or final representation.

5.9.1 Viewing the workspace

It is possible at any time to view directly any of the workspace views by digitising DIRECT VIEW X-Y, DIRECT VIEW Y-Z, or DIRECT VIEW X-Z. The desired view is then displayed so that it fills the screen.

Normally, when inputing in orthogonal mode, all views of the drawing are displayed simultaneously on the screen as they are constructed. If the user does not require all views to be displayed, by digitising DISABLE ORTHO VIEWS he can selectively SWITCH OFF any views. To re-enable display of a particular view the command ENABLE ORTHO VIEWS has the reverse effect. This facility is particularly useful for plotting if all 3 views are not required on a given drawing.

5.9.2 Perspective and isometric viewing

In pictorial mode the system is switched from perspective to isometric mode and vice versa by digitising the squares ISO MODE or PERSPECTIVE MOD respectively. The current mode is retained until it is changed by the user (the joystick function automatically switches to perspective mode).

In isometric mode it is possible for the user to specify an isometric origin on which the drawing is to be based or, more generally, to allow the system to place the origin automatically so that the drawing is placed centrally in the field of view.

Perspective projection is controlled chiefly by the joystick function, which manipulates the user's position of view to give the required projection. This is a relatively arbitrary process, but it is useful for obtaining rapid views from various positions. More accurate views can be obtained by specifying explicitly the position of the eye, direction of view, and focal length when required, although it is believed that the use of perspective mode is, in general, limited. Since a perspective view is not to a fixed scale, the system scales and moved it to fill the field of view, known as the perspective window, which may be reset by the user at any time. This 'fitting' process can result in alarming changes of scale if the user is panning around an object to obtain different views. It can be disabled by digitising the square USE OLD PERSPECTIVE FIT, and re-enabled by digitising REFIT PERSPECTIVE WINDOW.

5.9.3 Joystick

The joystick function allows the user to take full advantage of the flexibility of the CAD system; it allows him to view 3-D objects from any angle. An area of the menu 75 mm square is set aside for the joystick function. This mode is entered simply by digitising the square JOYSTICK on the menu. The system reverts automatically to perspective node, and the current angle of view is indicated by a small set of x,y,z axes displayed in the bottom left-hand corner of the screen. The user 'drives' to the required position of view by moving the cursor onto the joystick area; the system is made to pan by horizontal position, and to tilt by vertical position. When a satisfactory view is obtained the user enters the position and the system returns the corresponding projection.

When exiting from joystick this angle of view is retained until it is subsequently changed by either another call to joystick or by resetting any of the perspective parameters individually.

5.10 THE USER AREA OF THE MENU

Thirty squares on the menu have been reserved for user application or programs. Programs may be written in Fortran or MACRO 11 and interfaced into the draughting system. These programs are called via the user menu squares.

CHAPTER 6

Using the Draughting System

A simplified example of the use of the draughting system is described in this chapter. The application chosen is that of pipe network layouts. It illustrates the use of many of the facilities of the system such as the digitisation of data in three dimensions, the construction and use of macros, the use of isometrics, the annotation of drawings, and material take-off.

6.1 CONSTRUCTION OF MACROS

Macros can consist of two basic items. Firstly, a macro is a standard component such as a valve, flange, or pipe. Secondly, a macro can be made up of a section of a layout which is to be repeated.

It is necessary to build up a library of component type macros before the draughting system can be used to maximum effectiveness. The macro menu has 128 squares but it can be 'paged' to give a larger number of squares if necessary. Furthermore, each macro can be coded so that a single square may be used to call for example, a certain type of valve, and the actual size can be obtained from a predefined reference table based on the current input parameters, for example, pipeline diameter.

Take, for example, the valve shown in Fig. 6.1. It is first necessary to digitise the graphic symbol for the valve type. The graphical data representing the symbol must be in three dimensions, since the layout will have to be drawn as an isometric to allow possible clashes to be investigated.

The draughting system is first set up for use, as described in Section 5.1. The system automatically assumes a scale of 1:1 unless some other scale is specified. The drawing of the valve is attached to the digitiser.

When digitising a macro the user must first go through a short set-up procedure. All important parameters, such as input level and line type, are automatically given a default value on start-up, and it is up to the operator to change these to suit his needs. A typical drawing will probably be started by specifying an input scale, either manually or by digitising a line and specifying its length. This is followed by specifying the skew control to be applied to the drawing and then

Fig. 6.1 — A typical macro consisting of a valve.

the 3-D origin. The latter process involves digitising a position in 3-D and then specifying manually from the numeric input area the actual X, Y, and Z coordinates of this point. All subsequent points are scaled and translated relative to the predefined parameters. When digitising macros the input scale and orientation are not as important as when digitising a complete drawing, since the macro handler applies the necessary scale factors and rotations to the macro when it is recalled and before it is placed. When digitising, button 1 is used to input the coordinates of the beginning and end point of the lines, and button 2 is used to break at the end of line so that the cursor can be moved into another area.

Once the set-up procedure is complete the digitisation of the actual drawing may be commenced. Digitising data in 3-D is best done by selecting all the lines on a particular plane, for example the Z plane, and digitising one point of this data in XY followed by the corresponding point in XZ. The remaining points digitised in XY will remain on the Z plane already digitised until a new Z value is chosen.

The lengths of lines or distances are determined by using the TRAILING ORIGIN system in conjunction with the coordinate display unit on the digitiser. The operator starts by digitising the start point of a line and then moving the cursor, operating under CONTROL 90 if applicable, until the required line length or distance appears on the display; the digitising button is then pressed again. Line lengths may also be determined by digitising the dimension, using the numeric input under DRIVE MODE.

Each line is checked for position as it is digitised by viewing the screen. Any mistake can usually be spotted immediately and rectified by using the editing facilities. For example, if a line is input in error it may be removed by digitising

LINE EDIT on the menu followed by guiding the cursor on the screen until it is close to the line to be removed. When the cursor is within a predefined distance of the required line, the line is brightened and prompting messages are displayed. Any button will delete the line, or the search can be advanced by moving the cursor. Normal digitising is resumed by specifying EXIT on the menu.

Circular shapes, such as the valve handle, are digitised by using CIRCLE MODE on the SYMBOL menu. The circle centre and then any 2 points on the circumference are digitised to obtain the diameter and 3-D location.

When the macro is complete it is entered into the library, as described in Section 5.6, along with any relevant reference data, which will be used to manipulate the macro when it is recalled.

6.2 CONSTRUCTION OF LAYOUTS

To make CAD techniques really cost effective when applied to draughting, maximum use must be made of the macro facilities. Any standard items such as valves, filters, and pipe fittings will be stored in the macro library. However, any part of the layout which is subject to repetition may be digitised first into the macro library for subsequent use in the construction of the layout.

The procedure for construction a pipe network layout is to sketch the layout first with dimensions set down from a set of grid lines or datum. The rough drawing can then be placed on the workspace of the digitiser and turned into an accurate drawing. As an example consider the drawing of a simple pipe run shown in Fig. 6.2. We will show how the drawing can be accurately constructed and how documentation such as drawings of plans, elevations, and isometrics together with material schedules, can be produced.

The system set-up procedure is similar to that described for the construction of macros except that the drawing data is filed under a code and not a macro library square. Each file may be subdivided into a number of levels which are identified by number. This permits a complex drawing to be digitised in a number of discrete parts, and each part may be overlayed with another to make up the complete drawing. In the present example the various different pipelines have been assigned to different levels. All annotation is assigned to level 16, and a flag masked into the data indicates whether it applies to orthogonal or isometric views. Thus a drawing may have two completely different sets of annotation without them interfering with one another.

The first digitising stage is to set up the grid or datum lines, input scale, and reference point as described previously. Additionally, the user must specify the current pipeline type to enable correct recall of macros and a 'principal pipeline direction' to aid their positioning.

The next stage is to enter the pipe runs together with the standard components such as valves. The most common procedure is to pick macros from the library and place them in the correct position on the drawing, using the TRAILING ORIGIN facility. Complete macros can easily be manipulated by using the macro

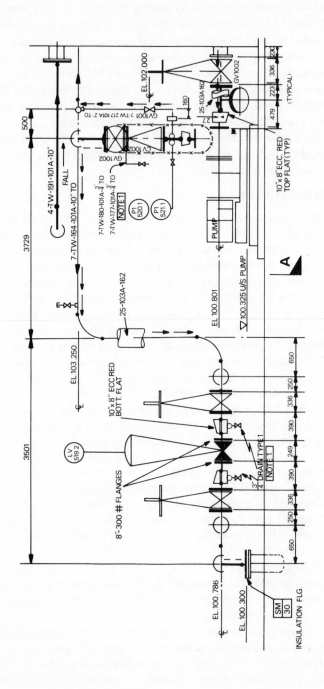

Fig. 6.2 – An example of a pipe run.

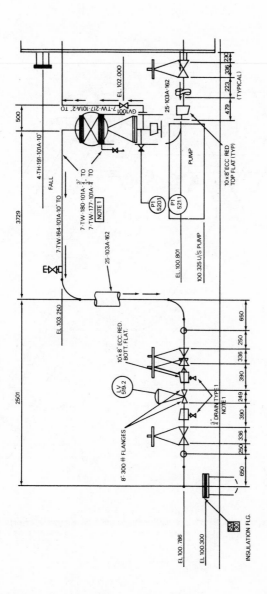

Fig. 6.3 – Final layout.

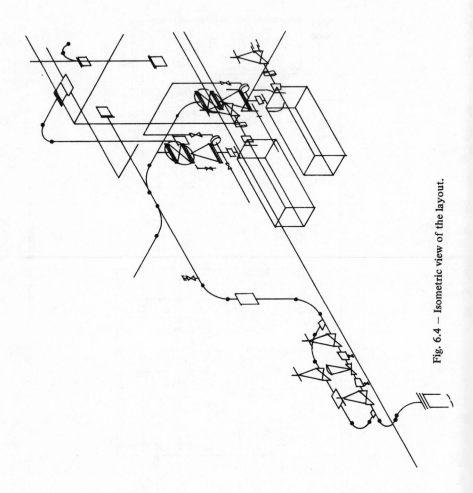

Fig. 6.4 – Isometric view of the layout.

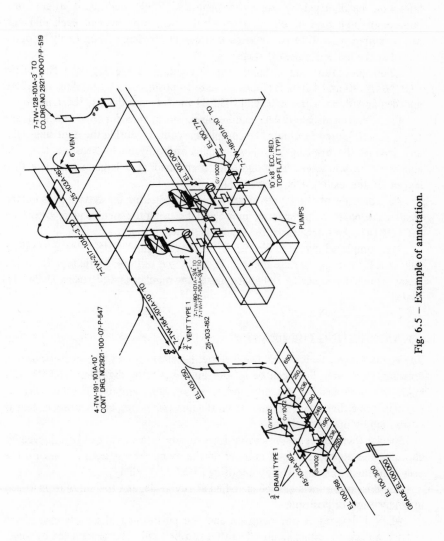

Fig. 6.5 — Example of annotation.

menu facilities. For example, if a macro has been placed in an incorrect position it can be deleted by the symbol editor and then repositioned by the macro handler.

Once all the components have been located in the workspace, the pipe centre lines may be digitised by using the menu and FIND facilities. If there are a number of pipe runs of differing diameters, then pipe runs of each diameter can be stored on a different level so that cumulative lengths can easily be determined at the material take-off stage.

When pipe runs are entered, the straight runs are digitised first. The CONTROL 90 and FIND facilities are used to position each run. Arcs and fillets are then added, using the ARC and FILLET modes from the SYMBOL menu.

Each level of digitised data can be displayed in turn for final checking, and levels superimposed to obtain the complete layout. If, during the digitising stage, any area of the drawing becomes complex and difficult to visualise, then that area may be windowed and digitising performed at a larger scale. The final layout of the example is shown in Fig. 6.3.

An isometric of the layout is produced by switching the system to isometric mode as opposed to perspective mode and digitising the menu square DISPLAY PICTORIAL. An isometric of the layout is shown in Fig. 6.4.

The completed layout is filed on DISK by digitising WORK SPACE TO FILE followed by the appropriate number. It can be put onto magnetic tape for permanent storage by selecting the appropriate tape number and digitising DUMP TO TAPE.

6.3 ANNOTATING THE DRAWING

The symbol menu contains all the facilities necessary for adding dimensions to a drawing. Dimension lines can be set out individually by digitising DIMENSION MANUAL followed by the appropriate annotation symbols. The lettering or numbers are either typed or input from the numeric menu, and a string of text is positioned by means of the cursor.

Special thicknesses of dimension lines or any other lines can be achieved by choosing the appropriate pen number on the menu. The system can inform the user as to the pen being used by digitising CURRENT PEN.

The dimension lines can be specified at any angle, and the annotation keyed in at any angle or position.

When a drawing is not complex and the positioning of annotations is not critical, automatic dimensioning facilities can be used. This is achieved by digitising DIMENSION AUTOMATIC followed by the digitisation of the 3 points where the dimension is required. It is necessary to determine the points accurately by using the FIND facilities. The computer will then calculate the distance between the points and construct the appropriate dimension lines, inserting the correct annotation. The computer will test for space to insert the dimensioning arrows and annotation. If the space is small, then the arrow heads and annotation are placed externally to the dimension lines.

It is often convenient to run off a plot of a layout or component drawing before annotating it. The plot, which is an accurate drawing, may then be annotated in freehand. The roughly annotated drawing is then placed on the workspace and annotated in the way described. This technique is useful for annotating large complex drawings since it allows the draughtsman time to carefully plan the annotation on the drawing. Performing the planning while using the system could be wasteful of computing time. Another advantage in using an accurate plot of the drawing during the annotation stage is that points on the drawing where dimensions are required can easily be located, since moving the cursor on the digitiser to the required point and using the FIND facility will ensure that a desired point is located quickly and precisely. An example of annotation is shown in Fig. 6.5. Any notes can be added to the drawing by typing strings of alpha-numerics of the desired length and positioning them with the digitising cursor.

Alpha-numeric data or dimension lines can be edited, where necessary, by using the symbol editor. Digitising SYMBOL EDIT, displaying the cursor over the symbol to be deleted, and pressing button 1, will remove the symbol.

6.4 PLOTTING THE DRAWING

High quality ink drawings are essential for accuracy in reading, especially if the drawings are being reproduced on a dye-line machine. The pens must therefore be properly maintained. Plotters can use a variety of pens: ball point, ink type, or even nylon tip with ink. If pens are left open to the atmosphere for long periods without use, problems usually result. The need to keep the pens clean and to follow the plotter manufacturer's instructions cannot be over-emphasised.

The GCADS system can produce data for a wide variety of plotters from large flat-bed machines such as the Kongsberg Kingmatic and CIL Envoy to the drum plotters by Calcomp or CIL. A data tape can be produced for any of these plotters by accessing an appropriate user menu square. The flat-bed plotter attached to the system is on-line to the computer and is simply accessed by digitising PLOT PICTORIAL or PLOT ORTHOGONAL on the Display and Manipulation of Data menu. The drawing produced on the plotter will be at full size unless the output scale is specified by using the instruction OUTPUT SCALE followed by the scale factor.

6.5 MATERIAL TAKE-OFF

The graphical draughting aspects of CAD are important, and in many applications it can be shown to be very much faster and more economic than manual techniques. However, once the draughting system is linked to analysis programs, the benefits become increasingly apparent. For example, the production of a material schedule is a simple matter and is effected by digitising the user command MATERIAL TAKE-OFF. The schedule of all the material used on the layout is then produced as a listing on the line printer.

CHAPTER 7

Transformation Systems

In chapters 4, 5, and 6 we have encountered various two-dimensional and three-dimensional transformations. These include scaling, rotation, translation, and general homogenous matrix transformations. The present chapter discusses how these transformations are made, and it is addressed primarily to those who may wish to write certain graphics software.

We can use hardware as well as software to perform transformations. There are a number of displays that use hardware to perform all the most common transforms, but these displays are expensive. Most displays require a level of software to perform transformations, and sometimes this software is located in a microcomputer display processor system.

The most important aspect of software-driven picture transformations is the speed of the transformation. In practice the calculation of transformations can be time-consuming and prove to be a bottleneck in the performance of a display system. To save time in calculating transformations it is often necessary to concatenate two or more transformations into a single transformation. The concatenating of transforms will be discussed.

7.1 DISPLAY

CAD involves displaying a large number of pictures, and much time can be consumed in converting structured data into display signals. The display file, as previously discussed, may be regarded as a table of instructions to be executed by the display processor. The display file, with hardware consisting of a digitiser and storage tube, may be the same as the workspace for a two-dimensional graphics system, since there is no need for a conventional display memory. In a three-dimensional system, intermediate files may be used to advantage, and this will be discussed later.

The display processor consists of a program and a number of subroutines built into a library which are loaded to execute the instructions contained in the display file. This is shown in flow chart form in Fig. 7.1. In the simplest data structure, the instructions are stored in $[I,X,Y]$ records with I representing the command code with respect to the display procedure and X, Y the data. Compiling of the graphics language is simple, for instructions are usually generated in a sequential manner. The instruction code I may have a wide range of meanings; for example:-

$I = 1$ drive beam from last position to position $[X, Y]$ with beam switched off.

$I = 2$ drive beam from last position to position $[X, Y]$ with beam switched on.

$I = 0$ stop display.

Efficient algorithms for converting display instructions to signals driving the CRT are particularly important for fast display of data.

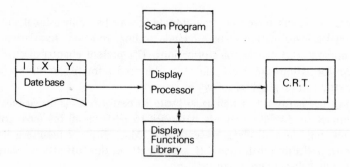

Fig. 7.1 — One-pass display processor.

7.2 WINDOWING AND CLIPPING

When it is necessary to examine in detail a part of a picture being displayed, a window may be placed around the desired part and the windowed area magnified to fill the whole screen. This involves scaling the data which lies within the window so that the window fills the entire screen. Data that lies outside the window must be eliminated so that only the data required for display is processed. This process is known as **clipping**. Some hardware devices have automatic scissoring in which the window and the display vectors may be larger than the display raster. For some systems, as with the present DSV system, hardware clipping is not available and software clipping must therefore be employed.

It is necessary to define two viewing areas, these being a viewport and a window. It is usual to make the viewport equal in size to the screen to take advantage of maximum screen area.

The limits of the window are determined by the coordinates on the bottom left-hand corner taken as [0,0] and the dimensions of the required frame. The window is set up by digitising the [0,0] coordinates and moving the cursor on the digitiser to the right until the window encloses the desired area. Once the window is defined, ,the data outside the window is clipped before scaling to the screen coordinates. This considerably reduces the amount of data before display signals are generated.

The Tektronix 611 storage tube used in the Imperial College system contained 1024 points horizontally by 737 points vertically. The windowing transformation is given by:

$$X_S = 1024 \frac{X_P - X_W}{D_X}$$

$$DX \qquad Y_S = 737 \frac{Y_P - Y_W}{D_Y}$$

where (X_W, Y_W) are the coordinates of the bottom left-hand corner of the window and D_X and D_Y are the frame length and height of the viewport defined on the table, as shown in Fig. 7.2.

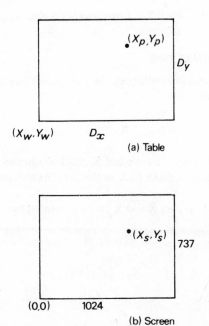

(a) Table

(b) Screen

Fig. 7.2 – Windowing transformation.

The clipping algorithm tests the position of lines in relation to the window. First it tests for the trivial cases where lines lie entirely within or without the window as shown in Fig. 7.3. If these tests are not satisfied, a line is assummed to intersect the window. The intersecting lines are trimmed so that the external points lie on the edge of the window.

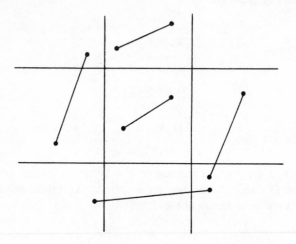

Fig. 7.3 – Trivial cases of clipping.

7.3 2-D TRANSFORMATIONS

Every picture manipulation performed in the system can be represented by a matrix transformation:

$$\mathbf{X}_T = \mathbf{X}_T$$

where $\mathbf{X}$ is a vector in the data file and $\mathbf{X}_T$ the transformed vector contained in a display file. $\mathbf{X}_T$ can be written back in the data file to represent a transformed new data source.

For two-dimensional data $\mathbf{X}$ and $\mathbf{X}_T$ are represented by:

$$\mathbf{X} = [x,y]$$

$$\mathbf{X}_T = [x'\ y']\ .$$

Thus T is a 2 x 2 matrix:

$$T = \begin{bmatrix} a_{11}\ a_{12} \\ a_{21}\ a_{22} \end{bmatrix}$$

where the terms a_{ij} can be defined for each transformation

$$[x\,y] = [x'\,y'] \quad \begin{bmatrix} a_{11}\,a_{12} \\ \\ a_{21}\,a_{22} \end{bmatrix}$$

T can be a single matrix or concatenated by multiplying two or more matrices. The coordinate values in X are normally relative to a reference point X_r. Therefore, in general:

$$\mathbf{X}_T = T\,\mathbf{X} + \mathbf{X}_r$$

where

$$T = T_1 \,*\, T_2 \,*\, T_3 \,*\, \ldots\ldots$$

The order of the individual matrices T_i determines the resultant transformation. However, using 2 x 2 matrices it is not always possible to concatenate the matrices. For example, to rotate a point $[x\,y]$ through an angle θ about an arbitrary point at a distance $[R_x\ R_y]$ from the origin, conventional operation in 2 x 2 matrices would give:

$$\text{translate } [x'\,y'] = [x\,y] - [R_x\,R_y]$$

so that $[R_x\,R_y]$ becomes the origin.

$$\text{Rotate } [x''\,y''] = [x'\,y'] \begin{bmatrix} \cos\theta & -\sin\theta \\ \sin\theta & \cos\theta \end{bmatrix}$$

and translate again to old origin

$$[x'''\,y'''] = [x''\,y''] + [R_x\,R_y] \ .$$

The transformation becomes:

$$\mathbf{X}''' = (\mathbf{X} - T)R + T \ .$$

The matrices cannot be combined into one single transformation, but this is possible by using homogeneous matrix representation [1], where each transformation in two dimensions is represented by a 3 x 3 matrix.

Although there is no practical advantage in storing data as homogenous coordinates, it is useful to perform transformations with 3 x 3 matrices. Adding a third element we obtain:

$$[x'\ y'\ 1] = [x\ y\ 1] \begin{bmatrix} a_{11} & a_{12} & 0 \\ a_{21} & a_{22} & 0 \\ a_{31} & a_{32} & 1 \end{bmatrix}.$$

The extra dimensional element represents a scaling factor and is always kept as unity because a single common scale is used which does not require to be constantly redefined and stored as data. This also reduces the data storage requirement by 25%.

The multiplication of the last column can be eliminated, reducing the computation of the matrix from 9 multiplications and 6 additions to 6 multiplications and 4 additions:

$$[x'\ y'] = [x\ y\ 1] \begin{bmatrix} a_{11} & a_{12} \\ a_{21} & a_{22} \\ a_{31} & a_{32} \end{bmatrix}.$$

The above translation and rotation become:

$$\text{translation } T = \begin{bmatrix} 1 & 0 & 0 \\ 0 & 1 & 0 \\ R_x & R_y & 1 \end{bmatrix}$$

$$\text{and rotation } R = \begin{bmatrix} \cos\theta & -\sin\theta & 0 \\ \sin\theta & \cos\theta & 0 \\ 0 & 0 & 1 \end{bmatrix}$$

$$\text{and } X = [x\ y\ 1].$$

The matrices can be combined as:

$$X''' = X(-T)\ RT$$

where

$$_5(-T)\ RT = \begin{bmatrix} \cos\theta & -\sin\theta & \\ \sin\theta & \cos\theta & \\ -R_x\ \cos\ \theta-R_y\ \sin\ \theta+R_x & R_x\ \sin\ \theta-R_y\ \cos\ \theta+R_y & 1 \end{bmatrix}$$

si

Again, the last column can be eliminated from the final matrix computation. It is, however, necessary for the three matrix operations.

7.4 THREE-DIMENSIONAL TRANSFORMATIONS

Three-dimensional transformations are performed on data by matrix operations similar to those described in the preceding section.

It is necessary to define a system of reference axes and adopt a convention for the direction of rotation before one can consider any transformations in three dimensions. The conventional right-handed reference set of orthogonal axes is used here and is shown in Fig. 7.4. The $x-y$ plane is chosen to correspond to any flat working surface such as the digitiser or the viewing screen. The z direction is always forward, toward the observer.

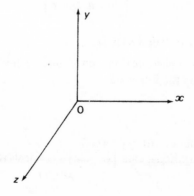

Fig. 7.4 — Right-hand reference set of axes.

The angle of rotation θ about any axis is taken to be positive the rotation is anticlockwise and negative when clockwise. Thus the rotation is said to be positive when in the sense of a right-handed corkscrew as shown in Fig. 7.5.

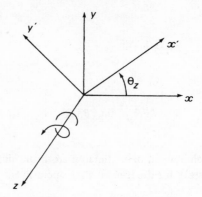

Fig. 7.5 – Rotation about z axis.

The rotation is given by the matrix:

$$\begin{bmatrix} \cos\theta & -\sin\theta \\ \sin o & \cos\theta \end{bmatrix}$$

with θ measured in the anticlockwise direction. The rotation refers to the set of axes on the plane of rotation. For rotation of objects in a space with a fixed set of reference axes the opposite applies and the matrix becomes:

$$\begin{bmatrix} \cos\theta & \sin\theta \\ -\sin\theta & \cos\theta \end{bmatrix}$$

7.5 LINEAR TRANSFORMATIONS

Three-dimensional transformations are similar to those in two dimensions. We shall concentrate on the following:

Scaling
Translation
Rotation (about one or more axes)
For three-dimensional data $[x, y, z]$ the transformation can be represented by:

$$\mathbf{X}' = \mathbf{X}T$$

where

$$\mathbf{X} = [x, y, z, 1] \text{ the original coordinates}$$

$$\mathbf{X}' = [x', y', z', 1] \text{ the transformed coordinates,}$$

and T can be represented by a single or concatenated 4 x 4 matrix:

$$T = \begin{bmatrix} r_{11} & r_{12} & r_{13} & 0 \\ r_{21} & r_{22} & r_{23} & 0 \\ r_{31} & r_{32} & r_{33} & 0 \\ t_1 & t_2 & t_3 & 1 \end{bmatrix}$$

where r_{ij} are the terms of the rotation matrices and t_j the translation offset.

The r_{ij} terms normally consist of a single or a combination of rotation matrices.

a)   Rotation about z axis (Fig. 7.5).

$$R_z = \begin{bmatrix} \cos \theta_z & -\sin \theta_z & 0 \\ \sin \theta_z & \cos \theta_z & 0 \\ 0 & 0 & 1 \end{bmatrix}$$

b)   Rotation about y-axis (Fig. 7.6)

$$R_y = \begin{bmatrix} \cos \theta_y & 0 & \sin \theta_y \\ 0 & 1 & 0 \\ -\sin \theta_y & 0 & \cos \theta_y \end{bmatrix}$$

c)   Rotation about x-axis (Fig. 7.7)

$$R_x = \begin{bmatrix} 1 & 0 & 0 \\ 0 & \cos \theta_x & -\sin \theta_x \\ 0 & \sin \theta_x & \cos \theta_x \end{bmatrix}.$$

R_x, R_y, R_z can be concatenated to give a general rotation matrix R. However, the order in which the individual matrices are combined will affect the results thereby achieved.

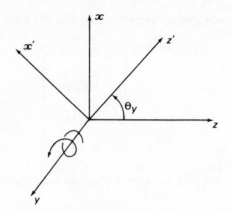

Fig. 7.6 – Rotation about y axis.

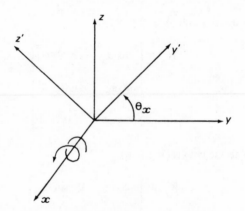

Fig. 7.7 – Rotation about x axis.

The scaling transformation is given by:

$$S = \begin{bmatrix} S_x & 0 & 0 \\ 0 & S_y & 0 \\ 0 & 0 & S_z \end{bmatrix}.$$

For equal values of S_i the scaling is linear, otherwise distortions will be introduced into the system.

These transformations allow data to be repositioned anywhere and in any orientation in space. The translation, rotation, and scaling effects can all be, combined into a single transformation matrix by concatenating the individual matrices. However, the same effects can also be achieved by varying the viewing parameters used to define the object-observer geometry in a perspective transformation.

The equivalent effects: pan, rotation, and scaling or zooming can be obtained by varying the observer's position and orientation relative to the object. This eye movement technique represents a more economical way, in terms of computing time, of obtaining movements from a static object. The data base remains unaltered, and only the viewing transformation, which is always required irrespective of the parameter values, is performed. The joystick function described in Section 7.9 uses this technique to provide very quick means of manipulating the display.

7.6 DISPLAY FILES FOR THREE-DIMENSIONAL DATA

The introduction of a third dimension increases the number of processes involved in converting data from a basic structure to drive a display terminal. There are two major stages: compilation and processing (Fig. 7.8).

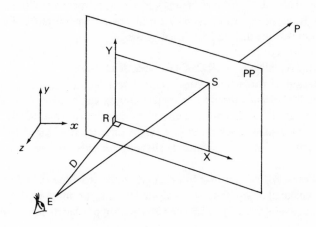

Fig. 7.8 – Projection on plane PP.

In the compilation stage the data file is first scanned by a viewing algorithm, an output routine which generates function calls to the appropriate subroutines (scaling, perspective, windowing, clipping). Then , in response to these function calls, the display compiler interprets the stored data and creates a display file. The cumulative effect of these two processes is to generate a second data structure from the first.

In the second stage, the display file, which contains simple commands for the display, is scanned by the display processor, which then generates the drive signals for the CRT. The instructions are simple $x-y$ drive signals, but the processor may incorporate some hardware character or symbol generation facilities.

These two stages are very similar, both involving a scanner and an interpreter, and can be combined into one, so that a set of data serves both purposes. With a storage tube a conventional memory for storing the display file is not essential, since the image is displayed only once. Because of the lack of selective erasure facilities the whole screen must be redisplayed to incorporate any changes. However, the use of a separate display file does have the advantage of making possible the use of the concept of segmented display files, which reduces the amount of compilation involved each time the picture is selectively modified. For example, in macro manipulation only the affected macro or modified segment is recompiled.

It is usual to work with two files, so that the data and display files are stored, for example, in FILE 1 and FILE 2 respectively. FILE 2 does not fall within the exact definition of a display file because it contains more than simple instructions for the display processor, and it also contains data that lies off the screen area. The main purpose of this file is to permit faster display at different window sizes without the need to repeat viewing transformations. It also permits interactive editing of the data base by a parallel search technique, and it is useful in visualisation test for three-dimensional viewing.

7.7 VISUALISATION OF THREE-DIMENSIONAL DATA

The presentation of three-dimensional graphic data on a plane is important since many people have great difficulty in understanding engineering drawings. When an object is rotated, for example, the viewer could be confused if the new view is not clearly presented. Such confusion might arise from a 'wire cage' drawing where it might be difficult to distinguish between lines representing the front and rear.

To help the CAD system user to interpret static two-dimensional projections of three-dimensional objects a range of visual aids may be adopted.

There are several ways in which the visualisation of objects can be improved:

a)   Perspective transformation.
b)   Brightness modulation.
c)   Hidden-line removal.
d)   Shading.
e)   Movement.

Techniques (a) and (b) mainly apply to wire-frame models, while (c) and (d) are associated with solid objects. These techniques are now discussed, since the use of the computer in this area is particularly significant.

a) Perspective transformation

[Perspective is one of the most valuable techniques for visualising objects. The transformation is often used when displaying three-dimensional data but it can also be used with two-dimensional data to represent a plane lying in three-dimensional space. Perspective is easily represented in mathematical form since it is based on elementary optics.]

Consider Fig. 7.8 where x, y, z represents the axes corresponding to the spatial cartesian coordinates. The system of reference axes used here is in the conventional right-handed reference directions. If the x–y plane is made to coincide with the screen, the positive z direction is out of the screen towards the observer.

For the purpose of the projection, a plane PP is defined with coordinates (X,Y). The projection maps the point $\bar{P}$ to a point $\bar{S}$ on the projection plane PP. The projection plane in this case corresponds to the viewing screen, with X to the right and Y upwards.

The observer's eye is situated at the point $\bar{E}$ $[E_x, E_y, E_z]$ and the line of sight makes angles a, β, γ with the spatial x, y, and z axes respectively. The angles a, β, γ can be specified by their direction cosines or by a point defined as the centre of attention, from which the direction cosines can be calculated. Normally the centre of attention is made to coincide with the origin or the volumetric centre of the object being observed.

Consider the plane of projection PP at a distance D from $\bar{E}$, normal to the line of sight. For an observer with the eye position at $\bar{E}$ $[E_x, E_y, E_z]$, direction of line of sight at angles $[a, \beta, \gamma]$ to the major axes, the point $\bar{P}$ $[P_x, P_y, P_z]$ in space is mapped onto the projection plane PP by the following relationships:

$$X = \frac{(S_x - R_x)\cos\gamma - (S_z - R_z)\cos a}{\sin\beta}$$

$$Y = \frac{(S_y - R_y)}{\sin\beta}$$

where $[X,Y]$ are the coordinates of the projected position of the point $\bar{P}$. $\bar{S}$ $[S_x, S_y, S_z]$ is the point of intersection of the line of sight on the projection plane and $\bar{R}$ $[R_x, R_y, R_z]$ the foot of the normal from the observer's eye to the plane.

[Isometric or orthogonal projections are often preferred to perspective drawings in engineering because measurements can be related more easily to the drawing. However, aesthetically the perspective view looks right and, since it is easy to produce from a CAD system, its use may well increase.]

b)   Brightness modulation

With this technique parts of the picture near to the observer are bright while those far away are dim. When this is required on a view an extra routine in the program is entered just before the vector generation, and this selects the required brightness levels for the display file as it is being constructed. When a picture has been constructed, the maximum and minimum z coordinates are noted. The z range is then divided into n regions where n are the visible brightness levels available in the display system. The picture is then displayed with the appropriate brightness level corresponding to its z region.

This technique is easy to implement and is very effective when displayed on the screen. It is difficult to obtain a hard copy version unless an electrostatic plotter is used.

c)   Hidden-line removal

If a complex three-dimensional drawing is fully displayed, then the large number of lines usually render the picture impossible to understand. The main problem is that the lines which are normally hidden by the object are all displayed, and this can lead to confusion. The hidden lines can be removed by the computer, but large amounts of computing time are usually required. Computation increases approximately as the square of the number of edges, therefore for moderately complex situations computation can become prohibitive on a small computer.

It is not easy to establish reliable algorithms to identify the lines to be removed. In general, the geometric calculations are straightforward if objects are convex polyhedra. But if the three-dimensional bodies are not rectilinear, the problem can be very difficult. A number of successful algorithms have been devised [2], [3] to perform hidden-line removal.

d)   Shading

Shading techniques have been developed extensively in work related to hidden-line removal, particularly at the University of Utah [4], [5] and at the CAD centre at Cambridge [6]. The technique is based on the recognition of distance and shape as a function of illumination.

The technique is similar to finite elements. The surface of a solid is divided into patches, and in regions of large curvature the patches are decreased in size. Each patch or element is then tested for visability and the degree of shading required. It should be remembered that hidden-line removal is a prerequisite for any shading algorithm.

The amount of shading required is determined by calculating the angle between the normal to the plane of the element and the vector direction of the propagation of the light. The normal vector can be calculated from the cross product of two vectors on the plane or from the equation of the plane. The angle of incidence is given by the dot product of the normal and the line of incidence. Brightness and visibility increase as the angle of incidence increases from 0 to 90°. For the visibility test only the sign of the dot product is required to determine whether the plane is facing the light. This is a preliminary test to eliminate all planes facing in the wrong direction. Gourand [5] used a method in which the intensity at the point where the elements meet is calculated. The intensity is then interpolated to provide smooth shading of the surface. For illumination a point or parallel beam source of light may be used. The surface can also be given a reflective index to make it shiny or dull or even transparent.

The output is obtained on devices that scan at different intensity levels, either by drawing a series of parallel lines or by overwriting. Results obtained by Parke, Gourand, and Newell are impressive, but they rely on powerful, special purpose hardware. The work at Imperial College by Yi on shading was based on the use of a Calcamp microfilm plotter.

e) Movement

Movement improves recognition of displayed objects. As an object is rotated or translated, ambiguities that arise from the superposition of points are eliminated and the geometrical properties of the object are revealed by the interaction of the points defining the object.

The storage tube is not really suitable for displaying movement since the whole picture must be erased before a new one is displayed. This can take time, so that real-time movement is not possible. However, it is possible to photograph a series of pictures from the screen of a storage tube so that they can be replayed to give real-time movement. This is particularly useful for a simulation exercise since it is an inexpensive method for obtaining a simple animated line test.

The use of movement in CAD will become increasingly important as new cheap refreshed displays are developed. The use of moving blue prints for the simulation of engineering dynamics is an exciting prospect for engineers. Movement, coupled with colour, it will be an extremely powerful tool in CAD.

f) Summary

The main drawbacks of most of these techniques are the large amount of computing power required and the need for specialised hardware in some cases. Some of these techniques place considerable strain on minicomputer based systems, and skilled programming is required for a viable system. Probably the most suitable technique for the minicomputer and storage tube combination is that of perspective.

The advent of microprocessors and new types of refreshed display is bringing near the time when three-dimensional graphics in colour with real-time movement will be a practical reality in low-cost CAD systems.

7.8 EYE CO-ORDINATES SYSTEM

In the preceding section a method whereby a perspective view of a three-dimensional object can be generated was described. The final drawing was made up of $[X,Y]$ points on the projection plane. However, it is not always desirable to represent an object as a flat drawing. Sometimes it is essential to preserve the depth information in order to determine the spatial properties of the solid, which would have been lost with a plane representation.

What is needed is a transformation which converts a three-dimensional object as viewed in perspective into another object which will give the same view when projected orthogonally (Fig. 7.9). The transformation, in effect, moves a local observer to infinity and distorts the object appropriately so that it still looks the same on the viewing screen. The transformation preserves all the spatial qualities of the object, so it is always possible to apply the three-

(a) perspective

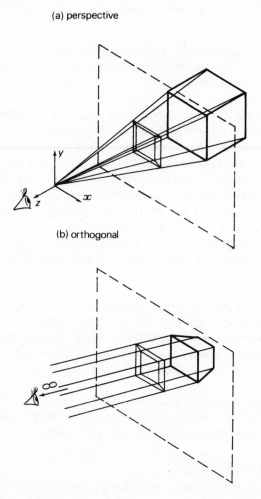

(b) orthogonal

Fig. 7.9 – Perspective and orthogonal projections.

dimensional perspective transformation before doing any visualisation computation like brightness modulation and hidden-line removal. It is much easier to perform hidden-line removal computation from an orthogonal projection than from a perspective projection.

Before the distortion transformation is applied it is necessary to define the object in terms of a local system of reference axes with a different orientation.

The new set of axes x', y', z' is taken with the origin at the eye position and with positive z' along the line of vision in the opposite direction (Fig. 7.10). x' is in the plane parrallel to the x–z plane, and y' is upwards. The angles that the line of sight makes with the x,y,z axes are given by their cosines $\cos\alpha$, $\cos\beta$, $\cos\gamma$ respectively .

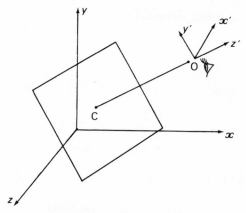

Fig. 7.10 – Eye coordinate system.

The transformation is a combination of a translation and a rotation.

$$X' = R (X - T)$$

$X = [x,y,z]$ point defined in the old coordinate sytem.

$X' = [x',y',z']$ point defined in the eye coordinate system.

T = translation vector given by $[E_x, E_y, E_z]$.

Matrix R is given by

$$\begin{bmatrix} \lambda_{x'x}, & \lambda_{x'y}, & \lambda_{x'z} \\ \lambda_{y'x}, & \lambda_{y'y}, & \lambda_{y'z} \\ \lambda_{z'x}, & \lambda_{z'y}, & \lambda_{z'z} \end{bmatrix}$$

where $\lambda_{x'x}$ is the direction cosine between x' and x, and so on.

From the above,

$$\lambda_{z'n} = \cos\alpha$$

$$\lambda_{z'y} = \cos\beta$$

$$\lambda_{z'z} = \cos\gamma$$

and x' is parallel to X
$\quad\ y'$ is parallel to Y.

$[X, Y]$ are the axes on the projection plane.

Now

$$\lambda_{x'x} = \cos\gamma/\sin\beta$$

$$\lambda_{x'y} = \theta$$

$$\lambda_{z'z} = -\cos\alpha/\sin\beta$$

and

$$\lambda_{y'x} = -\cos\alpha\ \cos\beta/\sin\beta$$

$$\lambda_{y'y} = \sin\beta$$

$$\lambda_{y'z} = -\cos\beta\ \cos\gamma/\sin\beta$$

thus

$$R = \begin{bmatrix} \dfrac{\cos\alpha}{\sin\beta} & 0 & \dfrac{-\cos\alpha}{\sin\beta} \\[3mm] \dfrac{-\cos\alpha\ \cos\beta}{\sin\beta} & \sin\beta & \dfrac{-\cos\beta\ \cos\gamma}{\sin\beta} \\[3mm] \cos\alpha & \cos\beta & \cos\gamma \end{bmatrix}.$$

Alternatively, R can be derived by regarding the transformation as a combination of two rotations (Fig. 7.11). The first is R_θ about the y axis in the x-z plane through angle θ anticlockwise (object rotation).

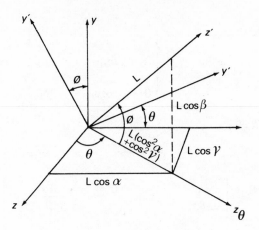

Fig. 7.11 – Rotation of axes in space.

Thus $R_\theta = \begin{bmatrix} \cos\theta & 0 & -\sin\theta \\ 0 & 1 & 0 \\ \sin\theta & 0 & \cos\theta \end{bmatrix}$.

From Fig. 7.11

$$R_\theta = \begin{bmatrix} \dfrac{\cos\gamma}{\sin\beta} & 0 & \dfrac{-\cos\alpha}{\sin\beta} \\ 0 & 1 & \cos\gamma \\ \dfrac{\cos\alpha}{\sin\beta} & 0 & \dfrac{\cos\gamma}{\sin\beta} \end{bmatrix}$$

where $\sin\beta = (\cos^2\alpha + \cos^2\gamma)^{\frac{1}{2}}$

and $R\phi = \begin{bmatrix} 1 & 0 & 0 \\ 0 & \cos\theta & -\sin\theta \\ 0 & \sin\phi & \cos\phi \end{bmatrix}$.

$$\begin{bmatrix} 1 & 0 & 0 \\ 0 & \sin\beta & -\cos\beta \\ 0 & \cos\beta & \sin\beta \end{bmatrix},$$

therefore $R = R_\phi R_\theta$.

As before,

$$R = \begin{bmatrix} \dfrac{\cos\gamma}{\sin\beta} & 0 & \dfrac{-\cos\alpha}{\sin\beta} \\[2ex] \dfrac{-\cos\alpha\ \cos\beta}{\sin\beta} & \sin\beta & \dfrac{-\cos\beta\ \cos\gamma}{\sin\beta} \\[2ex] \cos\alpha & \cos\beta & \cos\gamma \end{bmatrix}.$$

The above rotation matrix is valid for all directions of z' in Fig. 7.11 except when it is vertical. In either of the cases shown in Fig. 7.12 the direction cosines of the axis z' with respect to the reference axes can be determined by inspection, and the rotation matrix becomes

$$R = \begin{bmatrix} 1 & 0 & 0 \\ 0 & \cos\beta & 0 \\ 0 & 0 & -\cos\beta \end{bmatrix}.$$

The matrix is valid for both cases in Fig. 7.12. The direction cosine $\cos\beta$ is $+1$ for the first case and -1 for the second.

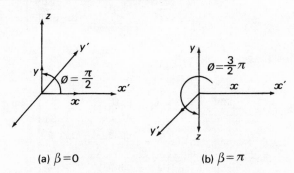

(a) $\beta = 0$ (b) $\beta = \pi$

Fig. 7.12 – Orientation of axes for a vertical line of sight.

In the new coordinate system the perspective projection becomes a simple linear relationship because the x'-y' plane is parallel to the projection plane.

From Fig. 7.13

$$[X\ Y\ Z] = [x'\ y'\ z'] \begin{bmatrix} D/_{z'} & 0 & 0 \\ 0 & D/_{z'} & 0 \\ 0 & 0 & 1 \end{bmatrix}$$

where D is the distance from the eye position to the projection plane. The z term is the depth coordinate and is stored in the z location of the display file.

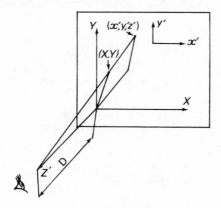

Fig. 7.13 – Linear perspective.

7.9 JOYSTICK FUNCTION

The variations in perspective parameters, as described earlier, allow a very convenient solution to a joystick controlled display.

The joystick function allows the operation of a joystick to be emulated on a patch on the menu. By pressing the pen button on the menu the user can control two functions: rotation and scale change or zoom. Rotations are possible about two axes: y axis and an axis on the x-y plane determined by the pan and tilt respectively (θ and ϕ in Fig. 7.14).

The eye position is defined by $[E, \theta, \phi]$ where

E — distance from centre of attention

θ — pan angle

ϕ — tilt angle.

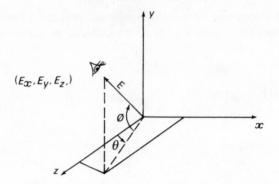

Fig. 7.14 – Eye position in terms of pan and tilt angle.

The coordinates of the eye $[E_x, E_y, E_z]$ with respect to the centre of attention are given by

$$E_x = E \, \cos\phi \, \sin\theta$$

$$E_y = E \, \sin\phi$$

$$E_z = E \, \cos\phi \, \cos\theta$$

and the direction cosines

$$\cos\alpha = E_x/E$$

$$\cos\beta = E_y/E$$

$$\cos\gamma = E_z/E \ .$$

Conversely

$$E = (E_x{}^2 + E_y{}^2)^{\frac{1}{2}}$$

$$\theta = \tan^{-1} \ (E_x/E_z)$$

$$\phi = \sin^{-1} \ (E_y/E) \ .$$

The observer is assumed to be always facing the origin of the coordinates system. For cases where the object is positioned so that it falls out of range, it can be repositioned by using the centering function, which automatically places the object with the centre of volume at the origin.

Movement of the object may be created by varying the observer viewpoint rather than the actual object position. With the centre of attention inside the object, the object is always visible on the screen. The coordinates of the object in the data file remain unchanged; only the perspective parameters are updated. Since the perspective transformation is always requested before display, and the time taken to update the paramerers is relatively insignificant, the resultant increase in display time is minimal.

Rotation is obtained by varying the two angles θ and ϕ. The relationships between θ and ϕ and the direction cosines $\cos\alpha$, $\cos\beta$, and $\cos\gamma$ are given (see Fig. 7.11) by:

$$\theta = \tan^{-1} \frac{\cos\alpha}{\cos\gamma}$$

$$\phi = \tan^{-1} \frac{\cos\beta}{\sqrt{(\cos^2\alpha + \cos^2\gamma)}};$$

$$\cos\alpha = \cos\phi \; \cos\theta$$

$$\cos\beta = \sin\phi$$

$$\cos\gamma = \cos\phi \; \cos\theta \; .$$

The zoom factor in the display is determined from the ratio between the distance of the plane of projection and the observer distance (D/E). It can be shown that magnification is proportional to this ratio. That is,

$$\frac{H'_2}{H'_1} = \frac{R_2}{R_1}$$

where H'_i is the image size for ratio $R_i = D_i/E_i$ and $D_i = E_i - P_i$, where P_i is the distance of the projection plane from the object. Any of the distances E, P, and D may be varied independently, but the effects achieved are different.

7.9.1 Distortion

Another ratio needs to be defined, the distortion, given by H'/H'' which is the ratio of the image sizes for two objects of the same size at a fixed distance apart.

a) Lens Zoom (constant P)

By fixing the plane position and varying the eye position we obtain a 'lens zoom' effect. The distortion varies with the magnification in the same way as the image obtained when changing the focal length of a zoom lens with the camera in a fixed position relative to the object.

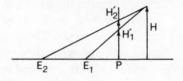

(a) constant P

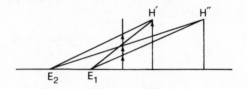

(b) distortion

(c) less distortion

Fig. 7.15 – Lens zoom.

Take Fig. 7.15(a). From similar triangles

$$\frac{H}{E_1} = \frac{H_1'}{D_1}$$

$$\frac{H'}{E_2} = \frac{H_2'}{D_2} \ .$$

Magnification, M, is given by

$$\frac{H_2'}{H_1'} = \frac{D_2}{D_1} \cdot \frac{E_1}{E_2} \ ,$$

therefore $M = \dfrac{R_2}{R_1} = \dfrac{D_2}{E_2} \cdot \dfrac{E_1}{D_2} \ .$

From Fig. 7.15(b)

$$\frac{H}{E_1} = \frac{H'_1}{D}$$

$$\frac{H}{E_1 + l} = \frac{H'_1}{D_1}$$

where H' and H'' are the image sizes of two segments of the same length H, distance l apart. E is the distance from the first object to the eye positions. The same relations apply for E_2, therefore distortion is given by:

$$\frac{H'}{H''} = \frac{E + l}{E} \, .$$

For a fixed P_1, given E_1 and D_1,

$$P = E_1 - D_1$$

$$R_1 = D_1/E_1$$

to obtain a magnification $M = \dfrac{R_2}{R_1} = \dfrac{D_2}{D_1} \cdot \dfrac{E_1}{E_2} \, .$$

The new eye position is given by

$$E_2 = \frac{P}{1 - R_1 . M}$$

and $D_2 = E_2 - P \, .$

b) Object Zoom (constant D)
The second alternative is to move the eye and the plane together by keeping D constant. The size change is given (Fig. 7.16(a)) by

$$\frac{H'_2}{H'_1} = \frac{E_1}{E_2} \, .$$

$$\text{Size } \alpha \, \frac{1}{E} \, .$$

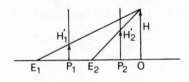

(a) constant D

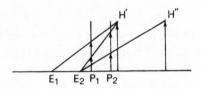

(b) distortion

(c) more distortion

Fig. 7.16 – Object zoom.

Again,

$$M = \frac{R_2}{R_1} = \frac{D}{E_2} \cdot \frac{E_1}{D} = \frac{E_1}{E_2} .$$

Therefore for a certain magnification M,

$$E_2 = E_1/M$$

$$D_2 = D_1 \quad .$$

The distortion given by $H_1'/H_1'' = (E_1 + l)/E_1$ actually increases when the magnification increases. The zoom has the same effect as closing in with a lens of fixed focal length on a fixed object, or moving the object towards the observer This is called an 'object zoom'.

c) Combined Zoom (constant E)
If the eye position is now fixed and the projection plane is moved we have a

'combined zoom'. This is equivalent to looking through a zoom lens and closing in as it is zoomed out. The distortion remains the same as the size increases.

$$\frac{H'}{H''} = \frac{E + l}{E}$$

is a constant now, because the eye position is fixed.

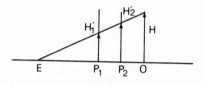

(a) constant E

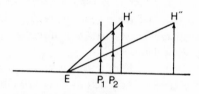

(b) distortion

(c) same distortion

Fig. 7.17 – Combined zoom.

From Fig. 7.17 a

$$M = \frac{H'_2}{H'_1} = \frac{D_2}{D_1} \text{ since } E \text{ is constant.}$$

Size αD.

Again, magnification $M = \dfrac{R_2}{R_1} = \dfrac{D_2}{D_1} \cdot \dfrac{E_1}{E_2} = \dfrac{D_2}{D_1}$.

For a combined zoom:

$$D_2 = D_1 \, M$$

$$E_2 = E_1 \quad .$$

The effects of the three types of zoom on a cube are depicted in Figs. 7.15(c). 7.16(c), 7.17(c). The three can be combined to give almost any type of zoom effect required.

For any eye-plane combination,

$$\text{magnification} \;\; \alpha \;\; R \;\; \alpha \;\; \frac{E - P}{E} \;\; \alpha \; \frac{D}{E} \; .$$

When performing a zoom effect it is essential that the projection plane P lies between the eye position and the object centre. That is D must always be positive, otherwise unpredictable results such as inversion, partial inversion, or very large scale magnification will occur.

The control of the parameters θ, ϕ, and R is via the joystick patch on the menu (Fig. 7.18). It is the same as the patch used for drive mode; it consists of a square 9 x 9 cm divided into 81 smaller squares, and a rectangle 6 x 9 cm divided into 9 rows. The square controls the tilt and pan angles, and the rectangle the zoom magnification. The angles are in increments of 1, 5, 15, and 45 degrees, or combinations of these. Rotations can be accumulated before display, which is activated by the 'enter' switch. The zoom zone is divided into three parts, each controlling a different type of zoom. The upper rows are for zoom-ins and the bottom ones for zoom-outs.

A point within the small squares must be digitised to obtain the desired orientation and magnification. The values of these are displayed on the screen.

The joystick control enables the user to select quickly the best orientation of an object for inspection. Because the display is linked point by point to the object data, it provides a convenient means of isolating data for construction and editing.

	COMBINED	LENS					TILT ANGLE						
	2.0						45°						
	1.5						15°						
	1.2						5°						
	1.1						1°						
	ENTER	ZOOM	45°	15°	5°	1°	Enter	−1°	−5°	−15°	−45°		
	0.75						−1°						
	0.5						−5°						
	0.2						−15°						
	0.1						−45°						

POSITION

PAN ANGLE

Fig. 7.18 – Joystick patch.

REFERENCES

[1] Roberts, L. C. Homogenous Matrix Representation and Manipulation of N-Dimensional Constructs, *Computer Display Review* 1969.

[2] Sutherland, I. E. Spinall, R. F. and Schumacker, R. A. A Characterisation of Ten Hidden-Surface Algorithms. *Computer Surveys* **6**, 1974.

[3] Watkins, G. S. A Real Time Visible Surface Algorithm. PhD thesis, Dept. of Electrical Engineering, University of Utah, 1970.

[4] Parke, F. I. Measuring Three-Dimensional Surfaces with a Two-Dimensional Data Tablet. *Computers and Graphics* No. 1., **1**, 1975.

[5] Gourand, H. Computer Display of Curved Surfaces. PhD thesis, Dept. of Electrical Engineering, University of Utah, 1971.

[6] Newell, R. G. The Visualisation of Three-Dimensional Shapes. *Proceedings of the Conference of Curved Surfaces in Engineering*, Cambridge, March 1972.

CHAPTER 8

Application of CAD Techniques to Finite Element Data Preparation

The finite element method is now widely used for the analysis of many engineering problems involving static, dynamic, and thermal stressing of structures. The technique is also used in other branches of engineering as well as reactor physics. In this chapter we shall deal with the application of finite elements to the stress analysis of structures. The technique of finite elements is not often easy to use, because of the problem of formulating the necessary input data for a finite element analysis program. The area of most difficulty lies in representing the geometry of the structure by suitable elements of regular shape. This is called the idealisation.

The input data for a finite element analysis program consists of the geometric idealisation, the material properties, and the loading and boundary conditions. The major problem, and a large portion of the input, is in the geometric representation of the structure by a suitable mesh. When this task is performed manually, it is time-consuming and subject to considerable errors since hundreds of punched cards are often required to describe an average-size problem. The human errors lead to abortive runs of the finite element analysis program, which can be expensive as well as time-consuming.

The data preparation stage is extremely tedious and time-consuming and consequently very expensive. Automating this process or mechanising it can greatly enhance the use of the technique.

8.1 AUTOMATIC MESH GENERATION

In many cases the user of a finite element program will reach a point where it becomes desirable to use an existing mesh generating system [1] or to write to one if necessary. Often it is difficult to write a program to generate a mesh for complex structures, and the user has to spend many hours drawing meshes and measuring the coordinates of each nodal point. In problems where the number of nodal points is very large, the user frequently has to spend many hours on data checking and rechecking by tedious graphical methods.

Using a CAD system in the data generation process has considerable advantages over most other methods. The user is able to see the element connections and position of each element directly on a display, as element generation is in progress. Also, if the system is interactive the user is able to change a mesh instantaneously to arrive at the best mesh arrangement to suit a particular problem by adding or deleting elements. The CAD sytem can also be used after the finite element analysis, to present data graphically so that the results can quickly be assessed.

If a CAD system is to be used in the generation and presentation of data in finite element analysis, it should ideally be capable of providing the following facilities:

Generation of mesh for two-dimensional or three-dimensional structures.

Ability to represent curved edges and surfaces.

Ability to control element density and generation of non-uniform meshes.

Facility for concentrating and grading the mesh over any region.

Speedy node and element numbering system which will lead to computational efficiency.

Facility to display idealisation.

Ability to rotate the idealisation from any desired angle.

Option to display any portion of the model.

On-line interactive modification of data for alternative idealisations.

Level of automatic mesh generation with manual override.

Preparation of input data for analysis programs.

The system should be user-oriented and easy to use with minimum of input, and economical with respect to both computer time and manual effort.

8.2 THE FINITE ELEMENT METHOD

The finite element method is now well established, and one of the best books on the subject is by Zienkiewicz [2]. The theory is based on an elastic structure or continuum being represented by many discrete components or elements interconnected at a finite number of nodal points situated on the element boundaries. The displacements of these nodal points are the basic unknown parameters of the problem.

A set of functions is chosen to uniquely define the state of displacement within each finite element in terms of its nodal displacements. Thus the displacement functions uniquely define the state of strain within an element in terms of the nodal displacements. These strains, with any initial strains and the constituent properties of the material, define the state of stress throughout the element.

A system of forces concentrated at the nodes and equilibrating the boundary stresses and any distributed loads is determined, resulting in a stiffness relationship of the form:

$$\{F\}^A = \{K\}^A \cdot \{U\}^A + \{F\}^A + \{F\}^A_{e0}$$

where

$\{F\}^A$ represents the nodal forces on element A;

$\{K\}^a$A is the element stiffness matrix;

$\{U\}^A$ represents the nodal displacements of element A;

$\{F\}_P^A$ represents the nodal forces required to balance any distributed loads acting on the element;

$\{F\}_{\epsilon 0}^A$ represents the nodal forces required to balance any initial strains such as may be caused by temperature change, if the nodes are not subject to any displacement.

It is not always easy to ensure that the chosen displacement functions will satisfy the requirement of displacement continuity between adjacent elements. Thus the compatibility condition on such lines may not be valid. Concentrating equivalent forces at the nodes ensures that equilibrium conditions are satisfied in the overall sense. Local violation of equilibrium conditions within each element will usually occur.

The choice of element shape and the form of the displacement functions will determine the accuracy of the finite element model.

It is also possible to define the stresses or internal reactions at any specified point or points of the element in terms of the nodal displacements:

$$\{\sigma\}^A = \{S\}^A \{U\}^A + \{\sigma\}^A + \{\sigma\}_{\epsilon 0}^A,$$

where

$\{S\}^A$ is the element stress matrix;

and $\{\sigma\}^A$ represents the nodal stresses for element A.

The last two terms are the stresses due to the distributed element loads and initial stresses when no nodal displacement occurs.

The stiffness matrix is related to the nodal forces, nodal displacements, and the type of element. Elements can take a wide variety of forms, the simplest being beam, triangular , and quadrilateral.

There are a variety of computer programs for solving the stiffness matrix, and hence deflections and stresses throughout a structure, for a wide range of elements. Some of the most widely used programs are NASTRAN [3] developed in the USA for the aerospace industry, and ASKA [4] developed in Stuttgart.

8.3 A GENERAL FINITE ELEMENT MESH GENERATING SYSTEM GFEMGS

A data preparation and results presentation system was developed by Ghassemi [5] at Imperial College for use with a variety of finite element programs. The system named GFEMGS is intended to work in conjunction with many of the modules of the draughting system GCADS described in Chapter 5. A user menu

GENERATE MESH FOR ANY QUADRANGLE	GENERATE CONCEN MESH QUADRANGLE	GENERATE MESH FOR PART DISC	GENERATE TRIANGLE QUADR-ANGLE	GENERATE HIGHER ORDER ELEMENT	CONNECT ISOPARA-METRIC ELEMENT
ADD ELEMENT TO MESH	DELETE ELEMENT FROM MESH	ADJUST NODES OF ELEMENT		DISPLAY INDIVI-DUAL NODES	DISPLAY ALL NODES
PROJECT ANY PLANE ADD 3D	DOUBLE X-Y PLANE PROJECT	3D COOR-DINATE FILE	3D ELEMENT CONNEC-TIONS	CHECK ANGLES & ELEMENT NUMBERS	PRINT ANGLES OF ANY TRIANGLE
INPUT BOUND-ARY CONDIT	ELEMENT WITH EXTERNAL FORCES		DISPLAY STIFF MATRIX	MINIMISE BAND-WIDTH	
DISPLAY RADIAL & AXIAL STRESS	DISPLAY CIRCUMF & SHEAR STRESS	CALCUL MAX SHEAR STRESS	DISPLAY STRESS IN 3D PICTURE	DISPLAY DEFORM-ATION	

Fig. 8.1 – USER menu for finite element data preparation and presentation system GFEMGS.

(Fig. 8.1) is introduced specifically to cope with functions required in the finite element data preparation or the presentation of results that are not in the GCADS facilities.

The system has two different modes of operation, one being automatic mesh generation and the other manual. Automatic mesh generation is used for two-dimensional and axisymmetric shapes. Meshes can be generated for a model by using a series of quadrilaterals with sides of different lengths. Each quadrilateral is automatically divided into a number of particular elements, such as triangular elements with three nodes, or quadrilaterals with four nodes. The user may subsequently transfer these elements to higher order elements such as triangular with six nodes, or isoparametric triangular elements with parabolic curved edges, or quadrilateral elements with eight or nine modes, and isoparametric quadrilateral elements with curved edges.

The user can generate a mesh within a quadrilateral so that it may be concentrated on any side or at any point within the quadrilateral. The system can also automatically generate meshes for circular discs, with or without holes, with meshes of various concentrations.

The GFEMGS system has a graphical checking system to check the coordinates and element connections, so that any error in the data will be shown to the user via the storage tube display or keyboard. For example, if after generating a mesh of a particular type of element a mistake has been made, then the check facility will identify any missing element on the display, or display any element with a suspicious shape such as obtuse-angled elements. The user is always guided through the data preparation, and actions are recommended at various stages of mesh generation.

The manual data preparation system makes extensive use of the draughting system. Each element is added to the model by using the FIND routine and the display. Thus a triangular mesh can be added by digitising two fixed points using FIND mode, and the third point of the triangle is moved by moving the cursor on the display until the desired position is obtained. An element can be deleted by digitising its nodes, using FIND, followed by an answer to a request to delete or ignore. The element is always displayed in bright-up form as soon as it has been identified, so that it is clear to the user that he has selected the desired element.

GFEMGS has an automatic nodal point numbering system which minimises the stiffness matrix bandwith. If the user has manually numbered the nodes then he may display the stiffness matrix on the display. If the bandwith is greater than a permitted value, then it may be renumbered by using the automatic system.

The user may identify the boundary conditions for a mesh directly from the digitiser. Nodes can be suppressed in any direction, or the loading specified. The load conditions are applied at the appropriate nodes as soon as the mesh has been generated and the elements which take the external forces have been identified.

Three-dimensional meshes are generated by a 2½-D technique. This involves the generation of a mesh on a plane, followed by a projection of this mesh onto a second plane parallel to the first. The projection can be smaller or larger than the original mesh. The process can be repeated many times until the desired size is obtained.

GFEMGS may be used to generate three-dimensional meshes by using any of the following elements:

Pentahedronal elements with six nodes or pentahedronal macro elements built of three tetrahedrons;

Hexahedronal elements with eight nodes or hexahedronal macro elements built of six tetrahedrons;

Triangular membrane elements in 3-space (three or six nodes); or quadrilateral plane membrane elements in 3-space (four, eight, or nine nodes).

GFEMGS has been run in conjuction with the finite element program ASKA which is available on the Imperial College CDC 6400 computer. Once the finite element data preparation stage is complete, the data is dumped to tape and transferred to the CDC 6400 for input to ASKA. On completion of the finite element calculation the results are transferred back to the CAD system for the presentation of results. GFEMGS allows the user to display deflections of the mesh graphically. Stresses are also shown graphically by a vector notation where the magnitude of the stress is proportional to the length of the vector, and the direction of the stress is indicated by the direction of the vector. Arrowheads at each end of the vector are used to indicate whether the stress is tensile or compressive. A second method used to display stresses takes the form of an oblique stress diagram which shows the stress projected away from a datum plane. Examples of results presentation are discussed in the next section.

8.4 AN APPLICATION OF GFEMGS

The use of the complete system is best illustrated by applying it to a practical example such as the standpipe from a nuclear reactor system which is shown in section in Fig. 8.2. The standpipe is subjected to an internal gas pressure of 39.8 bar when in use. The outside is subject to a water jacket at a pressure of 6.8 bar. The standpipe is also subjected to axial loads producing a pressure of 38.7 bar at the top and 234.9 bar at the base. The loading boundary conditions are shown in Fig. 8.3. Since the problem is axisymmetric, it was decided to use an axisymmetric ring element with a triangular cross-section (TRIAX3 in ASKA format) for the finite element mesh. The standpipe cross-section was divided into ten sections as shown in Fig. 8.4. The mesh for each section was generated automatically by digitising the four boundary points of each section and by specifying the number of nodes on each side of the sections and a mesh concentration factor.

The nodes in each section were numbered after the mesh was generated in a section. The mesh was generated automatically, starting from a specified corner, and the sequence of node numbering followed the sequence of the generation of the mesh.

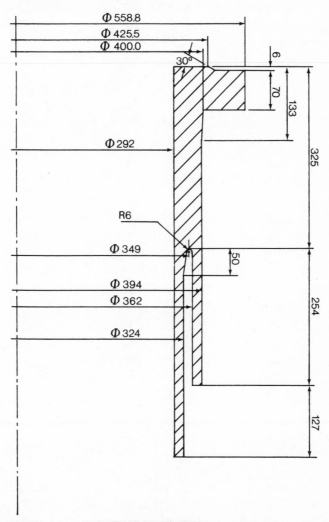

Fig. 8.2 – Standpipe.

One section was joined to another via a common set of nodes, and the numbering of a particular section was taken from the preceding section so that the common nodes are numbered only once. Each section is generated in a separate file to permit easy modifications to that section.

After the completion of mesh generation in all sections, the nodal points of each section were combined into a single file. The element connections were similarly dealt with.

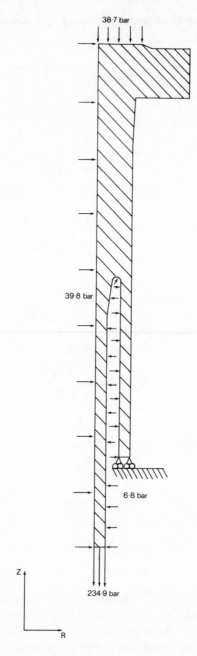

Fig. 8.3 — Boundary conditions.

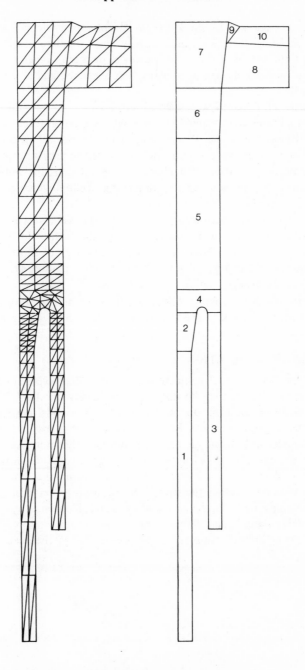

Fig. 8.4 – Mesh Generation for the Standpipe

The next stage was to check the nodal points and element connections, using the checking facilities which display the complete mesh. Errors are identified on the screen, and any obtuse angles between elements are indicated. The order of nodal and element numbers was examined and any serious errors were rectified.

The type of freedom was selected for each node, and the appropriate boundary conditions were applied at nodes where external forces were present.

The stiffness matrix was then displayed to examine its general pattern. If the bandwith is large owing to poor numbering of the matrix, where there is a large difference between any two adjacent node numbers, the whole mesh can be renumbered by an automatic renumbering facility which minimises the bandwith.

Finally the mesh coordinates, node numbers, element connections and numbers were transferred to magnetic tape for transference to the CDC 6400 main frame for input to ASKA. A record was also output to the line printer.

The mesh shown in Fig. 8.4 was generated in less than 40 minutes. A total of 438 statements were generated.

The finite element program ASKA was then run and the results output to tape for transference back to the CAD system. The ASKA run time was 120.5 seconds.

8.5 RESULTS PRESENTATION

The results consist of nodal point displacements, radial, axial, circumferential, and shear stresses. They may be displayed graphically on the CRT display or plotter or in numerical form on the line printer. The nodal displacements for the standpipe are shown in Fig. 8.5. The displacements may be magnified by using any desired multiplying factor. The stresses are shown in Figs. 8.6 to 8.13. Each set of stresses is displayed in two forms, one in vector form and the other as an oblique diagram. In the case of the vector form, the directions of the arrowheads indicate whether the stresses are compressive or tensile, and in the case of the oblique diagram, the nature of the stress is indicated by the direction of the projection away from a datum plane.

Finally, two further examples of meshes generated by the system are shown in Fig. 8.14.

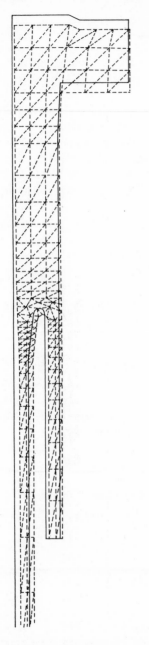

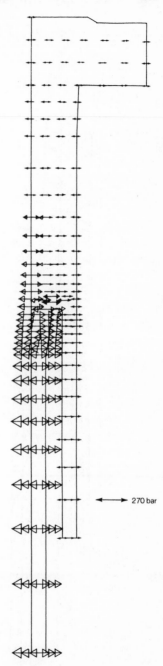

270 bar

Fig. 8.5 – Nodal point displacement.   Fig. 8.6 – Circumfoential stress.

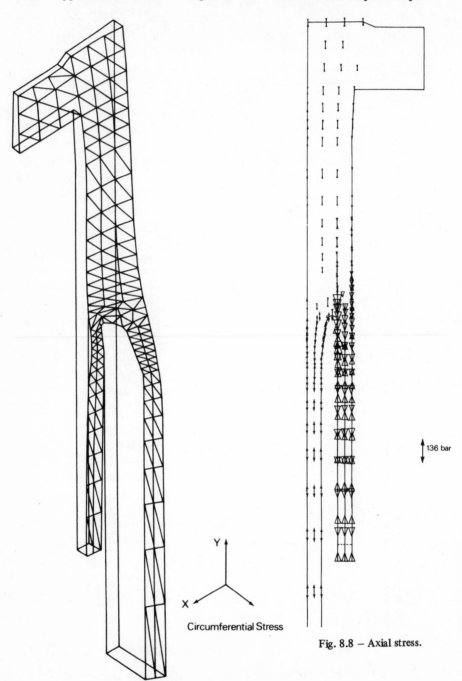

Circumferential Stress

Fig. 8.7 – Circumfoential stress.

136 bar

Fig. 8.8 – Axial stress.

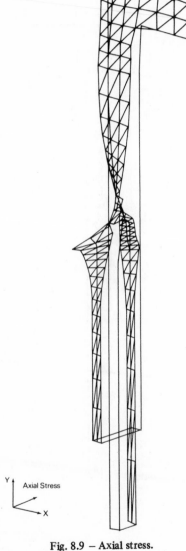

Y
↑ Axial Stress
→ X

Fig. 8.9 – Axial stress.

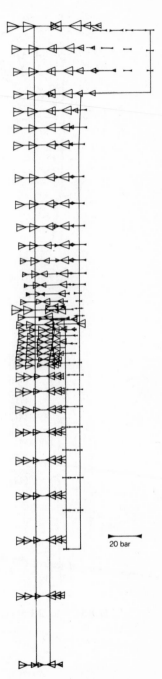

20 bar

Fig. 8.10 – Radial stress.

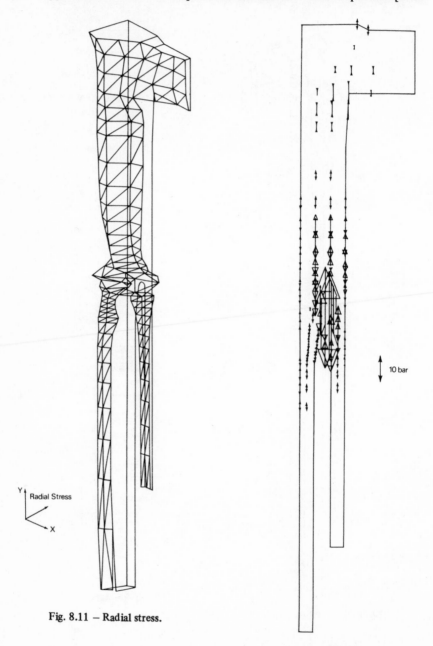

Y
Radial Stress
X

Fig. 8.11 – Radial stress.

10 bar

Fig. 8.12 – Shear stress.

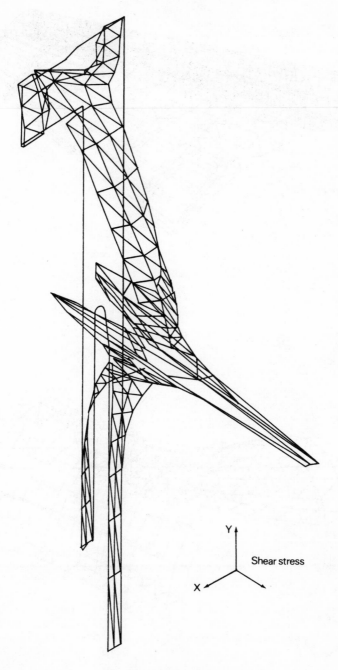

Fig. 8.13 – Shear stress.

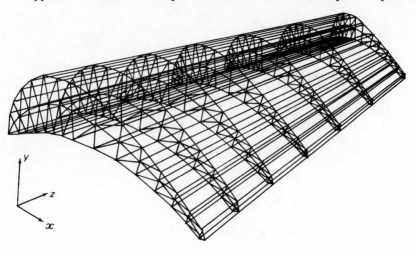

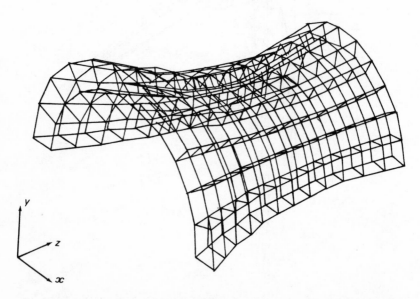

Fig. 8.14 – Meshes generated by the CAD system for an aircraft wing and a cooling tower.

REFERENCES

[1] Zienkiewicz, O. C. & Phillips, D. V. An automatic mesh generation scheme for plane and curved surfaces by isoparametric co-ordinates. *Int. J. Numer. Math. Engng.*, **3**, 519-528, 1971.

[2] Zienkiewicz, O. C. *The finite element method in engineering science.* McGraw-Hill, London, 1971.

[3] McCormick, C. W. (ed) *The NASTRAN User's Manual.* National Aeronautic and Space Administration.

[4] *ASKA linear static analysis user's reference manual.* ISD - report no. 73. Stuttgart, 1971 (ASKA UM. 202).

[5] Ghassemi, F. *Data presentation for finite element analysis.* PhD thesis, University of London, 1978.

CHAPTER 9

The Application of Interactive CAD Techniques to Machining Processes

9.1 INTRODUCTION

The use of computer systems in the engineering industry is widespread for both the design and manufacture of components. The fields of manufacture and design are invariably separate, but it is a natural step to decrease the gap between the drawing board and the manufactured item. Also, if manufacture and design are more closely linked, it is inevitable that the amount of effort required to proceed through design to manufacture will be reduced. The way to achieve this is to ensure that work carried out in the design process is not needlessly repeated during manufacture. If designs are produced by means of CAD techniques, so that the geometry is in a computerised form, this data can be presented to the manufacturing department instead of, or with, the drawings. The success of this procedure depends on the type of manufacturing techniques available and the machines used.

Many companies are turning to computer-aided manufacturing (CAM) techniques, mainly in the form of numerically controlled (NC) machines, in order to provide greater flexibility in batch production.

At present, integrated CAD/CAM systems are mainly used in problems involving surface definition, particularly in the aircraft industry. These problems are often very complex, and they produce such large quantities of data for both design and manufacture that computerised methods are essential.

Computer graphics facilities and the availability of supporting software have given designers the tools they need to produce computerised geometrics, and several successful packages using interactive graphics have been developed. One of these is GPP [1] (Graphic Post Programmer) – used extensively by the British Aerospace Military Aircraft Division. Similar interactive graphics software tailored to the needs of the motor industry has been developed by British Leyland at Cowley. IBM have marketed the FMILL and APTLOFT systems which are non-interactive surface machining packages. Probably the best known and most widely used software for this type of application is the CADAM system developed by Lockheed.

Less sophisticated industries than the aircraft industry are using NC machines and are becoming interested in CAD/CAM systems. It is in this wider context that research at Imperial College has been conducted. The aim of this work has been to integrate CAD and CAM by linking the existing computer aided draughting facilities with a manufacturing suite of programs geared to NC machine technology.

When an item has to be manufactured the first decision to be made is whether to use numerical control or more conventional methods. If the object is to be produced in sufficient quantities or is complex, then numerical control methods may be used. In the case of NC, a part-programmer will be given an engineering drawing of the required object and will produce a part program manuscript. This involves the coding of the geometry of the item (a repetition of work carried out during the design process) followed by the coding of statements to describe the tool motions required to carry out machining.

The manuscript, when complete, will be processed by an NC processor, and if no errors are found, a control tape will be produced. The coding and testing of the part-program is time- consuming and expensive. In the case of a large batch this initial time and expense can be justified. However, for a one-off job or small-batch production this may not be so.

The integrated CAD/CAM system developed at Imperial College by Craig [2] is based on the similarity of the output from the design process (defining the geometry of the design) and the input for the manufacturing process. This means that the design and manufacturing software can be completely separate but can share a common data base. The design process is made as interactive as possible to give the designer maximum flexibility; but once the design is frozen, the software for the manufacturing process will take the design data with the minimum of man interaction, and will produce a part program for submission to the APT system [3].

The term APT stands for 'Automatically Programmed Tools' and refers to both a language and a computer program. The APT language describes the sequence of operations to be performed by an NC machine.

The APT part-program is a series of instructions in pidgin English which describes the tool and its motions. The part-program is input to the main APT program, which runs on a large computer. The program converts the simple instructions of the part-program into a set of numerical commands by performing the necessary computation. These commands are then processed by a part-processor program which converts the numerical commands into a form suitable for a particular machine. The final-form numerical commands are produced on punched tape the control (CL) tape. The sequence of events is shown in Fig. 9.1.

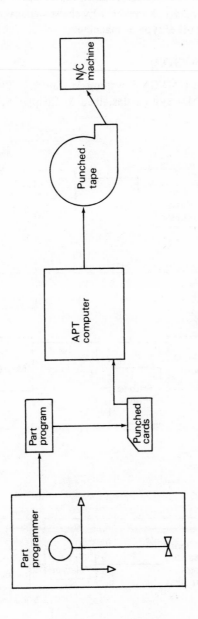

Fig. 9.1

The CAD/CAM system described in this chapter was developed for machining moulds for castings and is based on the use of a triple-axis NC milling machine. There is no reason, however, why the system cannot be used to provide input to APT for any other type of machine.

9.2 THE CAD/CAM SYSTEM

The hardware of the CAD/CAM systems is shown in Fig. 9.2. The CAD part is based on the GCADS system described in Chapter 5, consisting of a mini-

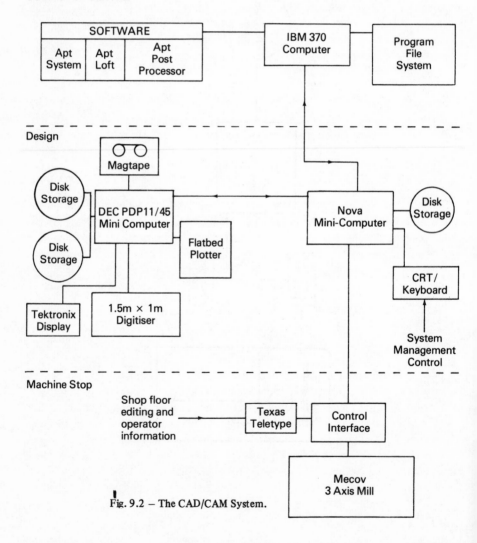

Fig. 9.2 – The CAD/CAM System.

computer with a digitiser and Tektronix display. Hard copy is via a flat-bed plotter.

The manufacturing part of the system is based on a NOVA minicomputer controlling a MECOV triple-axis mill.

The NOVA is linked to an IBM 370 mainframe where the APT processor resides.

At present a part-program generated by the CAD system for APT is output onto punched paper tape. It is then read onto a file on the disc of the NOVA. It is then transferred to the IBM 370 for processing and post-processing. The data is then sent back to the NOVA for implementation on the MECOV mill.

It is possible to terminate operations on the mill at any time during the execution of a block of instructions and to insert new instructions. It must be emphasised that there are a number of points around the system where users can interact with the software. Some of these interactions will be of a management nature in operating the system.

Only one CAD work station and one machine tool are shown in the CAD/CAM system. It is not difficult to have more work stations on the CAD system or more machine tools on the NOVA. At present, data transfer between the CAD system and NC system is via punched paper tape. It is planned to eliminate this system and replace it with a direct link.

9.3 SYSTEM SOFTWARE

The system software is based on the general draughting system described in Chapter 5. The manufacturing programs are called from the USER section of the menu. The interface between design and manufacture is a file known as the workspace file. It is based on fixed length records; each record consists of coordinates (x,y,z) and five numbers which qualify the coordinate. These numbers are packed into one integer word by the computer, and where necessary a further record can be used to describe a coordinate.

9.3.1 CAD Software

The CAD software has already been discussed in previous chapters, but certain aspects of it which relate to the machine tool application will now be amplified. We are concerned with solids which can be represented graphically in a number of ways. Orthogonal projection is the well-established method of engineering drawing, with at least two views which are plane areas (more views may make the visualisation simpler). It is relatively simple to obtain three-dimensional coordinates from such drawings [4]. The input facilities of the system obviate the need for the drawings to be of standard engineering drawing office quality: rough sketches will suffice.

We have seen how the two-dimensional surface of the digitiser is divided into areas corresponding to an X-Y, Y-Z, or Z-X plane in order to digitise in three dimensions. It is possible to use more than three planes, but usually only three are needed.

It is necessary to digitise the same coordinate in two planes in order to obtain a three-dimensional coordinate. However, the trailing origin facility obviates the need to do this in many cases, since if a series of points lie on one plane, say at a particular z level, then the x,y coordinates can be digitised until a new z coordinate is necessary. This approach makes the digitising of cross-section and contour data a very simple process. All the standard draughting facilities such as FIND, DRIVE, and CONTROL 90 give the user the power necessary to handle 3-D digitising.

In many cases, it is useful to digitise with respect to a reference grid which may be used as a reference skeleton of the object to be defined. A three-dimensional grid is set up in the following way. A grid origin is defined by digitising a point in a two-dimensional view; then grid increments are set up independently in any of the orthogonal directions. For example, the cursor may be constrained to move on a particular grid in the X and Y directions, whereas its movement could be left entirely arbitrary in the Z direction. The grid points can be displayed on the storage tube, and ten grid files are available for storage and retrieval of commonly used grid configurations.

Points are entered off the grid by using DRIVE mode, and it is also possible to disable the grid temporarily to input some arbitrary points and then to enable it again.

When the user wishes to define a component which is clearly 2½-dimensional, then all that is necessary is to digitise one of the required plane faces of the component and enter DOUBLING mode. The required plane is indicated to the program, and the thickness is defined. The program automatically generates the data for the solid and stores it in the workspace file.

Components which have one or more axes of symmetry can be generated by using the mirroring facilities of the system which allow data to be mirrored about any plane defined by the user.

Some of the more difficult work in manufacture is the machining of curved surfaces. Consequently the specification of curved surfaces in CAD is important. Curves on a surface can be defined in two ways. The first method produces an approximation to a curved surface and is performed by using a method called continuous digitising. Here the user traces the outline of the curve with the digitising cursor, and data points are stored automatically by the computer at short intervals along the curve. This is a speedy way of producing contour and cross-section data, but it is only as accurate as the user's skill in tracing the curve. If data points are known only at sparse intervals along a required curve, the second method should be used. Here, the points on the curve are defined by the user, either by digitising or using DRIVE mode, and a cubic spline is passed through the points, using CURVEFIT. This technique ensures that a curve is fitted to the specified points so that continuity of curvature and slope is maintained at all spans [5].

There are surfaces such as moulds for castings which do not easily lend themselves to definition in mathematical terms. The worst cases could be sculpted surfaces of an artistic nature. These surfaces tend to be described by a series of sections together with contours where necessary. A set of programs modules has been developed to cope with such surfaces; it will produce, from the input data, a surface which is smooth and which contains all the data points. The technique is based on the use of the facilities already described. The method of obtaining a surface requires the specification of a number of sections from given data.

There are three methods available for specifying sections. In the first method the user is asked to specify two points on a line. This line is then used to intersect with the data in the X-Y plane of the drawing and to return the $x,y,$ and z coordinates of the intersection points in an ordered form.

In the second method two points on a line are defined and a third point is digitised to indicate the bounds on a sweep. The program finds the intersection for a series of lines parallel to the one digitised within the bounds of the sweep.

In the third method a centre is defined, a number of lines at various angles are passed through this centre, and the intersections are found and stored as before.

When a sufficient number of cross-sections have been defined, the spline blending modules are initiated. The lines as defined previously are then tested for intersection in the X-Y plane. When intersections are found between two lines, a cubic spline is generated by the set of points on each line, and the Z ordinate at the point of intersection is found for each line. These ordinates are averaged and are taken as being points on the two lines; they are added to the data describing those sections. Unsatisfactory points are eliminated at this stage. When this process has ended, the output is the raw data plus a set of cubic splines which intersect in three dimensions with the raw data and themselves. The next stage is to generate a set of splines in the X and Y directions which contain the raw data and the generated splines. The final output is the raw data (as input) plus a family of splines which can readily be used to machine the surface.

Examples of a surface defined by raw data, raw data with blended splines, raw data with a family of orthogonal splines, and tool movements are shown in Figs. 9.3 to 9.7. It is possible to edit mistakes in the data.

Annotated drawings can be produced, and filing facilities are available for archival storage of completed designs.

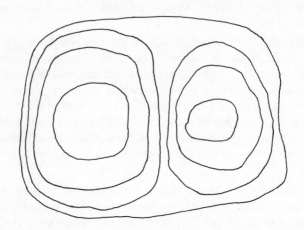

Fig. 9.3 — The original contour data describing the required surface.

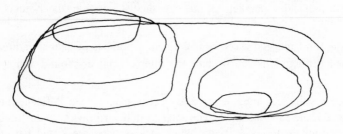

Fig. 9.4 — A perspective view of the above data.

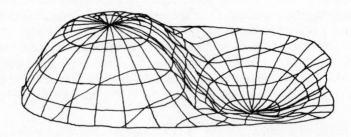

Fig. 9.5 — The set of blended splines defined by the user to describe the surface.

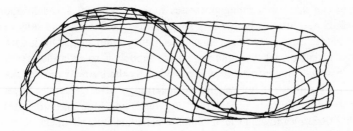

Fig. 9.6 – Here the data from the previous figure has been used to generate the family of orthogonal splines which are used to define the surface.

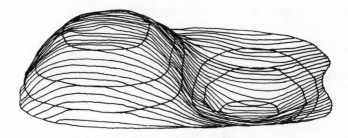

Fig. 9.7 – A plot of the computer simulated tool movement over the surface which has been calculated by the CAM program using the design data from Fig. 9.5.

9.3.2 CAM Software

The CAM program modules are entered after the definition and acceptance of the component designed by the user. In the design process it was deemed essential that the user should be in complete control of the machine, and input was made as flexible as possible. In the manufacturing area the NC modules play the leading part and operate on the workspace data. Interaction between the user and machine is still necessary but is minimised since the user only supplies information to the modules after a request from them. The CAM programs initially inspect the workspace data and attempt to split the data up into various machining sections. For example, to produce a sculpted facia panel it would be necessary to machine the basic shape of the panel, which would probably be rectangular, and then to machine the sculpted surface as a secondary operation. This discrimination of data by the NC programs is aided by the fact that it is possible to determine which of the CAD modules was used to generate the data in the workspace. Where the machine is unable to discriminate between surfaces, the user is called upon to interact and help to specify the surfaces more clearly.

When the basic machining processes have been decided, the component is inspected to determine whether any rough cutting is necessary to remove excess material before generating the accurate machining commands. This is determined to a large extent on whether the mould (or model) is to be milled from a solid billet or casting. This information is provided by the user. At present the user is requested to supply all information concerning cutter sizes, feed rates, and spindle speeds; but it is hoped that the user will soon have the option of specifying these or allowing the machine to calculate suitable ones.

The information supplied by the user and the data in the workspace allow the computer to automatically generate a file which is a digital representation of the part-program manuscript. In this file certain numbers are used which correspond to APT geometry words, and points in the workspace are referred to by a numerical label corresponding to their record number.

On completion of this process any or all of the tool movements over the component can be displayed on the storage tube, or drawings can be produced. All these displays are based on the data in the file which is created by the CAM programs (see Fig. 9.8).

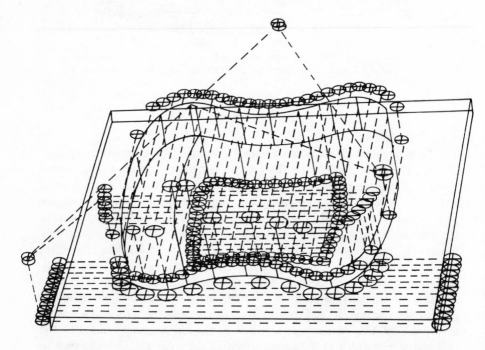

Fig. 9.8 – Typical tool paths which can be displayed on the Tektronix storage display or output in drawing form on the flatbed plotter. The tool paths are superimposed on the geometric description of the object to be machined so that checking can be easily performed.

At this stage the production of the symbolic part-program manuscript can be carried out, using a translating program. This part-program can be stored in any one of a hundred possible files on the computer's disc.

APT can be programmed in two basic ways. The most commonly used, and the one used by the CAM programs where possible, is to describe the tool motion over a component in terms of the relationship of the tool to three surfaces, the part drive, and check surfaces. Where this is inconvenient, as when producing tool movements from the data defining the 'artistic' surfaces described earlier, the CAM programs use APT in the straight point-to-point fashion in which the tool head is moved from one discrete point to another.

The USER section of the menu for the CAM system is shown in Fig. 9.9.

9.4 THE FUTURE OF CAD/CAM

The operation of the CAD/CAM system described shows that it is possible to almost automatically derive the necessary information to machine an object by working from a very simple data base which is little more than a wire-cage description of the object.

The time savings in programming for the machine tool mean that small-batch production and one-off manufacture can be brought into the range of the NC machine. At present the system is unlikely to produce as efficient a part-program as that produced by an experienced part-programmer, who will eliminate all unnecessary tool movements. However, an experienced part-programmer might use the present CAD/CAM system to modify the computer-generated tool paths to obtain a more efficient program. The required modifications could be speedily performed by using the interactive properties of the system.

Recent work has resulted in a CAD/CAM system in which the numerically controlled machine tool is driven directly from the minicomputer without the need for the APT system. The overall cost savings in bypassing the large computer where the APT system normally resides, are very significant and point to a future in which CAD/CAM could be very cost-effective. The development of microcomputer systems will have a tremendous impact on CAM since the machine tool of the future will have great power in machining complex shapes with relatively simple input instructions.

	CURVE-FIT						
INPUT SECTIONS OR CONTOURS	DISPLAY	STORE	DEFINE SPECIAL SURFACE	SPLINE BLEND	GENERATE ORTHO-SPLINES	DEFINE CUTTER START POINT	DEFINE TOOL CHANGE POINT
POINT TO POINT MOTIONS	DEFINE CUTTER	DEFINE FEED RATE	DEFINE RAPID FEED RATE	ENTER/EXIT MAN-MODE	ZERO MAN FILE	ELGAMILL POST-PROCESSOR	OUTPUT CL-TAPE
MACHINE COMMANDS	PLOT MAN FILE	DISPLAY MAN FILE	ZERO INT FILE	OFFSET FILE TO INT FILE	STORE TO INT FILE	ADD Z OFFSET TO INT FILE	GET STATUS OF INT FILE
DISPLAY INT FILE	INT FILE TO MAN FILE	INPUT SURFACE	DISPLAY SURFACE	INT FILE TO 3-D	AREA CLEAR	DRAW ANGLE	ROUGH CUT
INTERFERENCE CHECK	INPUT BOUNDARY	DISPLAY BOUNDARY	GENERATE OFFSET BOUNDARY	DISPLAY OFFSET BOUNDARY	OFFSET BOUNDARY TO STORE	DISPLAY STORE	DELETE STORE

Fig. 9.9 – Special USER menu for CAM.

REFERENCES

[1] Coles, W. A. APT at British Aerospace. CAM-I International Seminar, 1975.

[2] Craig, D. P., Besant, C. B., Hsiung, C. Y. & Williams, G. M. J. *The use of interactive CAD techniques and NC machining in mould design and manufacture.* 18th International Machine Tool Design and Research Conference. Imperial College, 1977.

[3] APT Part Programming, IIT Research Institute. McGraw-Hill, 1967.

[4] Yi, C., Besant, C. B. and Jebb, A. *Three dimension digitising techniques for Computer Aided Animation. Proceedings of an International Conference on Computers in Engineering and Building Design.* CAD 76. Imperial College, 1976.

[5] Nutbourne, A. W. Curve fitting by a Sequence of Cubic Polynomials. *J. of CAD.* Vol. 1, No. 4, 1969.

CHAPTER 10

The Implications in Industry

10.1 THE SAVINGS IN COSTS

The savings in costs which occur to industry as a result of using CAD techniques are not always apparent and are at times difficult to assess. Many people who try to assess the cost-effectiveness of CAD tend to concentrate on the use of CAD in draughting, which in practice is only a small and specialized application of CAD. There are many areas where automated draughting systems show considerable savings in cost and time over traditional draughting methods. Such fields tend to be in structural engineering, architecture, and electronics. This is not always so, because in mechanical engineering the advantages of using automated draughting systems are often insignificant both in terms of times and costs.

Let us consider present-day costs in manufacturing industry. The material and direct labour costs, although they are significant, are not usually as high as the overhead costs which can often amount to 60%. The overheads are related to the costs in carrying an idea right through to a product on the production line, ready for manufacture. It is here that CAD and CAM can make a big impact in saving time and costs. The economic efficiency of CAD is frequently contrived rather as a result of having done the job more quickly than having it done more cheaply. This is done in a variety of ways: for example, the rapid completion of a large project can have a large effect on the cash flow with such an effect that the speediest through-put of work from the same number of personnel will reduce the job costs, hence the time spent on estimating and tendering can be significantly reduced.

10.2 INTEGRATION

The benefits of CAD arise from the improved integration within an organisation. Staff within that organisation, by use of CAD, tend to work with a common database, with the benefit that information created in one department need not be duplicated in another department but is accessed as required. For example, if a component requiring NC machining is designed by using CAD techniques, then the production engineer can also use the geometric data describing the component, which had been created by the designer, in order to produce a control tape for the NC machine via an APT or similar program. It is also possible to integrate production control and material scheduling more closely within the design process. This is because the designer can have access to data and can define the types and availability of machines and materials which are necessary for manufacture. This can avoid time-wasting which might result from the unavailability of the required machine tools or materials.

Integration on a wide scale is clearly apparent in certain construction companies using CAD methods where, for example, architects and structural engineers are using common data-bases for draughting requirements. The storing of information and access to that information are benefits derived from the use of CAD techniques. Storing information in computer data-banks can be cost-effective. Generally, permanent files of data are stored on magnetic tape which is cheap, although accessibility is not fast. Data which requires immediate access is stored on magnetic disc, and these discs are usually of the exchangeable cartridge variety. If many departments within a firm can become linked into computer data-base systems, then the generation of larger amounts of documentation can be reduced. For example, if NC machines are used in turbine blade manufacture, then complex drawings describing the turbine blade shape are no longer required, because the manufacturing data can be passed over to the shop floor in the form of a CL tape. The shop floor can always gain access to the original data by a remote graphics terminal if any verification of the data is required.

The main virtue of CAD is in its interactive approach to design. The close association of graphics and analytical software can result in an extremely powerful tool giving rapid design coupled with reduced costs whereby the design can be 'frozen' at an early stage. The interactiveness of a CAD system can be used with great success to guide a user through a difficult piece of analysis. Consequently, complex analytical techniques, such as finite element stress analysis, can be made more readily available and easy to use without recourse to specialist engineers.

10.3 FUTURE TECHNICAL ADVANCES

The future of CAD and CAM will be greatly enhanced by new communication technology and by microprocessors. Improved communication techniques will result in more communication between men and computers. This will permit engineers to access powerful computing techniques from a terminal which can be far removed from the computer. Furthermore, the terminal could be as small as a conventional pocket calculator.

Microprocessors are already making a huge impact on CAD and CAM. A new breed of machines ranging from digitisers, plotters, and computer terminals to machine tools is being constructed with a built-in level of 'intelligence' by using a microcomputer. This permits a much greater flexibility in the use of such machines because they can be programmed to perform a much wider range of tasks than those with a fixed control system. For example, it is desirable to make a plotter look like a teletype so that if it is connected to a computer then simple instructions may be given to it; and if it contains sufficient local intelligence, it can draw complicated shapes from these simple instructions. Thus, the greater the amount of local intelligence that can be built into CAD peripherals, then so much the greater is the number of work stations which can share one mini-computer. This results in a better utilisation of equipment.

Display technology is also improving, with refresh displays becoming competitive with the storage tube display. The refresh display can bring the possibility of movement to pictures, which in turn can lead to simulation exercises. This is therefore a very important field in engineering which can reduce the need for experimental work with hardware and which can therefore save time and costs. For example, CL tapes for NC machine tools can be verified by simulation techniques.

10.4 THE TRADE UNION VIEW

Perhaps the attitude of trade unions to the use of CAD and CAM techniques will be more significant than all the new technical developments. We have seen how unions and management disputes in the printing industry can result in the closure of a national newspaper over the proposed introduction of computerised typesetting and printing equipment. Many companies are finding it increasingly difficult to introduce CAD techniques in their drawing offices. Unions, such as the Amalgamated Union of Engineering Workers (Technical and Supervisory Section), have looked more closely at the effects of the use of CAD techniques on their members and have come to the conclusion that such methods should not be introduced into a company without full consultations with the unions.

The unions see many potential dangers in CAD and CAM. These dangers are well described by Cooley [1]. The unions see a continuing fragmentation of jobs, just as the designer's job in the 1930s became broken down into a series of specialised jobs in the 1940s, with stress-men needed to perform calculations, metallurgists to select materials, tribologists to determine the form of lubrication, and the draughtsmen to perform the drawing.

The unions think that the employers will wish to exploit their members more and more as high capital cost equipment is used in design and manufacture. They see a higher degree of specialisation resulting in less job satisfaction because fewer will be able to see the panoramic picture of the complete job. It has been suggested that, as personnel become highly specialised, then employers will wish to keep them in the same job until that job disappears when the men will be 'scrapped' along with the machines.

It is claimed that talk in industry about dedicated machines and computers is commonplace, and this means that men will become a dedicated part of a system similar to a component in a machine. Even graduates will become more specialised, needing degrees in electronics, control or heavy engineering, rather

than in electrical engineering. It is thought that CAD, in some cases, takes specialisation to a very high level so that personnel may only have a short working life.
The unions further claim that students are being trained as industrial fodder
rather than being trained to think.

Probably the most dangerous aspect seen by many is the high burn-up rate in
staff which the intensive use of CAD facilities can bring about.

For example, one American company quotes 95% of a designer's time as
being spent in searching for information, and only 5% in the actual design decision
making. The introduction of computer graphics can eliminate much of the
routine reference work and can intensify decision making by over 1900 per cent.
As a consequence the stress put on the man can become very great. It is claimed
that, because the job requires a greater intensity of work, then only personnel
within a certain age bracket will be recruited into those jobs, and that as staff
grow older they may be subjected to career de-escalation with a consequent
lowering of their status. Redundancies often show that it is the older men who
are being eliminated.

It is the change in job structures and the 'de-skilling' of traditional labour
that most worries unions. For example, the setting of machines once performed
by skilled men on the shop floor is now moving to the drawing offices where,
with the use of CAD equipment, control tapes are generated for NC machine
tools.

The unions also point out that apart from all the negative effects that CAD
can bring to its members, the union itself can be in a strong position because its
members are becoming increasingly important to the support of high capitalised
equipment, and that therefore employers are vulnerable to strikes from a few
members of staff.

Some trade union members are aware of trends in automation in countries
such as Japan and the USA. They realise that competitors from these countries
are using computer controlled machinery to great effect. A typical example is the
use of computer controlled robots in the Japanese car and tractor plants. The
increase in efficiency that accrues from the use of such machinery will result in
products that are far more competitive than similar products by less sophisticated
techniques involving a high labour content. It is now being realised by some that
new techniques in design and production must be accepted in the long term, or
otherwise jobs will disappear to overseas competitors.

10.5 THE EMPLOYERS' VIEW

The employers' side of the CAD picture is not quite so clear as that presented by
the unions. Undoubtedly, there have been a number of companies which have
employed the use of high-cost CAD equipment and which have gone to extreme
lengths to maximise the return on the capital invested. Rating and work study
methods were introduced to increase productivity in some cases.

The latest trends in CAD and CAM are moving away from that equipment which has a high cost to user ratio. Minicomputers and, more recently, microprocessors are resulting in the development of CAD and CAM equipment whose cost is low enough to make 'idle time' economically tolerable. In other words, people are recognising that a designer must have time to think, or otherwise he will be forced into errors by excessive pressure of work.

Many companies who introduce CAD systems into their drawing offices do so only after careful planning and consultation with their staff. Where automatic draughting systems have been placed in the drawing office, the draughtsmen have often welcomed its introduction, because much of the tedious work associated with draughting is performed by the machine. This saving leaves more time for the designers to think and to plan the work. A helpful policy is the sharing of one CAD work station among a number of draughtsmen or designers. Each man can then plan his work and go to the CAD system for short periods when sketches and ideas can be worked up into proper designs. Acceptance of this approach gains a considerable increase in productivity without subjecting the workers to any severe stress: indeed, job satisfaction usually increases, because the drudgery of the job is removed and the results of a day's work can be that much more impressive. Very often, designers will add software of their own to the CAD system for the performance of additional tasks on the machine.

The use of CAD may not necessarily cause more specialisation, but quite the reverse. There are considerable moves away from specialisation because of the inflexibilities that result. CAD helps this move from specialisation because a designer with a good all round training can call on specialist programs, such as finite element stress analysis programs, which can be built into the CAD system.

University establishments, such as Imperial College, are moving away from the policy of producing highly specialised graduates for industry and are instead offering three-year and four-year 'total technology' engineering degrees. These courses are planned to give graduates a wide base of engineering subjects blended together with as much as 25% of course content in the humanities; that is, in management, economics, sociology and languages. During their studies, students spend time working in industry, and often work overseas as an extension of experience gained in their own sponsoring firms. The students are encouraged to read engineering rather than be taught and 'spoon-fed'. The wider awareness gained from the reading of books and literature will broaden their minds and outlook and enrich their actual experience. The object of the university is to turn out graduates who can think, so that each can make his own useful contribution throughout his entire working life. It is not the aim to produce graduates for a specific job which might last for only a few years.

Computing is playing an increasingly important role in educating engineers, because their skills can be widened by giving them a greater breadth of engineering knowledge.

The manner in which CAD and CAM are used in the future depends on every person concerned with engineering, with each one taking a positive attitude to its introduction and use, at the same time having real concern for the people who work in the engineering world, or indeed in any other field of human activity, because this problem will not stop at engineering but will affect society everywhere.

REFERENCES

[1] Cooley, M. *Computer Aided Design – Its Nature and Implications.* AUEW (TASS) 1972.

Index